Series/Number 07-094

LOGLINEAR MODELS WITH LATENT VARIABLES

JACQUES A. HAGENAARS
Tilburg University

SAGE PUBLICATIONS
International Educational and Professional Publisher
Newbury Park London New Delhi

For information address:

SAGE Publications, Inc.
2455 Teller Road
Newbury Park, California 91320
E-mail: order@sagepub.com

SAGE Publications Ltd.
6 Bonhill Street
London EC2A 4PU
United Kingdom

SAGE Publications India Pvt. Ltd.
M-32 Market
Greater Kailash I
New Delhi 110 048 India

Printed in the United States of America

Library of Congress Catalog Card No. 89-043409

Hagenaars, Jacques A.
Loglinear models with latent variables / Jacques A. Hagenaars.
p. cm.—(A Sage university papers series. Quantitative applications in the social sciences; 07-094)
Includes bibliographical references.
ISBN 0-8039-4310-5 (pbk.)
1. Loglinear models. 2. Latent variables. I. Title.
II. Series: Sage university papers series. Quantitative applications in the social sciences; 07-094.
QA278.H333 1993 93-21639
519.5′35—dc20

03 04 10 9 8 7 6 5

Sage Production Editor: Susan McElroy

When citing a university paper, please use the proper form. Remember to cite the current Sage University Paper series title and include the paper number. One of the following formats can be adapted (depending on the style manual used):

(1) HAGENAARS, JACQUES A. (1993) Loglinear Models With Latent Variables. Sage University Paper series on Quantitative Applications in the Social Sciences, 07-094. Newbury Park, CA: Sage.

OR

(2) Hagenaars, J. A. (1993). *Loglinear models with latent variables* (Sage University Paper series on Quantitative Applications in the Social Sciences, series no. 07-094). Newbury Park, CA: Sage.

CONTENTS

SERIES EDITOR'S INTRODUCTION

Contingency table analysis has recently experienced numerous major advances, almost all which have been explicated in this series. Some of the topics addressed include loglinear modeling (DeMaris, 1992; Knoke & Burke, 1980), latent class analysis (McCutcheon, 1987), causal modeling (Asher, 1976; Davis, 1986), factor analysis (Kim & Mueller, 1978a), and LISREL (Long, 1983b). These distinct methods are built upon in this synthesizing, yet innovative, contribution by Professor Hagenaars.

After an introduction, the basics of loglinear modeling are presented in Chapter 2. First, elemental notions, such as that of an "odds ratio" and a "logit" (i.e., the logarithm of the odds ratio) are defined. Then, saturated and nonsaturated loglinear models are developed. Finally, systems of logit equations are offered, in a "modified path analysis."

In Chapter 3, the complication of latent (i.e., unobserved) categorical variables enters, using an example from survey data on political action. The interpretation of the latent variable depends on the connections between the latent and manifest (i.e., observed) variables, in a fashion similar to factor analysis. Eventually, "external" variables as well are incorporated into causal models with latent variables, using a "modified LISREL approach" (see Chapters 4 and 5). One feature of a modified LISREL approach is that causality must be in one direction. In the last substantive chapter, the potential of the method is extended to longitudinal data, through exploration of panel data on American and Dutch voting intentions.

Partly because loglinear modeling with latent variables is on the methodological frontier, at least for practitioners, computer routines are not available in standard statistical packages such as SPSS or SAS. However, in Appendix A Professor Hagenaars provides an extensive listing of sources where programs are obtainable. After reading his concise, step-by-step exposition of the technique, a reader should be capable of using one of these programs to carry out thoughtful, sophisticated contingency table analysis.

—*Michael S. Lewis-Beck*
Series Editor

LOGLINEAR MODELS WITH LATENT VARIABLES

JACQUES A. HAGENAARS
Tilburg University

1. INTRODUCTION

> During the past decade a revolution in contingency table analysis has swept through the social sciences, casting aside most of the older forms for determining relationships among variables measured at discrete levels. (Knoke & Burke, 1980, p. 7)

With these opening lines of their 1980 Sage University Paper *Log-Linear Models,* Knoke and Burke described how during the 1970s loglinear modeling became the dominant form of categorical data analysis. Since then, loglinear modeling has strengthened its position as more and more social investigators have mastered the principles of loglinear analysis and have fruitfully applied loglinear models in their research (see, for examples from the area of social mobility research, Erikson & Goldthorpe, 1992; Hout, 1989; Sobel, Hout, & Duncan, 1985). Moreover, the standard loglinear model itself has been and still is being expanded in several directions (Agresti, 1990). One of the major new developments will be discussed in this monograph: loglinear models with latent variables (applications of which are provided by, among others, Clogg, 1979; Hagenaars, 1990; Hagenaars & Halman, 1989).

Latent variable models come in many varieties. No doubt, the best-known latent variable model is the classical factor analysis model (Harman, 1976; Kim & Mueller, 1978a, 1978b). The empirical starting point for factor analysis is a correlation or covariance matrix rendering

AUTHOR'S NOTE: *I would like to thank Elsbeth Kuijsters for her competent and enthusiastic assistance with the data analyses; J. Scott Long and an anonymous reviewer for their thorough, detailed, and very helpful comments on the original manuscript; Marianne Sanders for her skillful correction of the many common errors of a Dutch writer; and Michael Lewis-Beck for his substantial and substantive editorial support. WORC—the Work and Organization Research Centre of the Faculty of Social Sciences at Tilburg University—generously provided financial support.*

the relationships among the observed continuous variables measured at interval or ratio level. In factor analysis, one tries to find directly measured, continuous latent variables or factors that explain the relationships among the directly observed manifest variables. The latent variables found must have a theoretically meaningful interpretation based upon the factor loadings, that is, in agreement with the strength and direction of the relationships between the factors and the manifest variables.

If the directly observed variables are discrete, measured only at nominal or ordinal level, a categorical data analogue to factor analysis is provided by latent class analysis (Andersen, 1990; Langeheine & Rost, 1988; McCutcheon, 1987). The starting point is a multivariate frequency table rendering the relationships among the directly observed categorical variables. Analogous to factor analysis, in latent class analysis one tries to find theoretically meaningful discrete latent variables, each having two or more latent categories or classes, that explain the relations among the categorical manifest variables.

One recent major breakthrough in the analysis of continuous data has been the bringing together of factor analysis and ordinary regression analysis into one powerful overall model, the best-known variant of which is the LISREL model (Bollen, 1989; Long, 1983a, 1983b). In much the same vein, ordinary loglinear analysis and latent class analysis can be united into a general loglinear model with latent variables leading to what could be called, in line with Goodman's terminology, a modified LISREL approach (Goodman, 1972, 1973; Haberman, 1979, chap. 10; Hagenaars, 1988b, 1990, chap. 3). As will be exemplified throughout this monograph, the potential of this modified LISREL model for the analysis of categorical data is similar to what the LISREL model has to offer for the analysis of continuous data. It is possible to test and correct the categorical measurements for various forms of unreliability and invalidity and to estimate the true relationships among the categorical characteristics; in longitudinal studies, true changes may be separated from apparent changes caused by measurement error.

The general principles of loglinear modeling with latent variables are explained in Chapter 4. This explanation is preceded by an introduction into ordinary loglinear modeling (Chapter 2) and standard latent class analysis (Chapter 3). The explicit use of a loglinear model with latent variables as a causal model is the topic of Chapter 5. Because latent variable models are especially relevant for the analysis of change, Chapter 6 is devoted to the loglinear analysis of categorical longitudinal data. In Chapter 7, a brief evaluation of the loglinear approach with latent variables is presented, summarizing its main strengths and limitations along with some important new developments. A list of readily available computer programs is presented in Appendix A; Appendix B is a technical appendix on the EM-algorithm for obtaining the estimated expected frequencies for the class of models discussed in this monograph.

2. THE LOGLINEAR MODEL

The purpose of this chapter is to provide an introduction into the basics of ordinary loglinear modeling. Those readers who have taken an elementary loglinear class but whose knowledge has become rusty will benefit most. Those who are completely unfamiliar with loglinear models should first read the Knoke and Burke (1980) contribution to this series or Reynolds (1977). Readers who want to know more about loglinear models are referred to the intermediate-level books by Fienberg (1980), Hagenaars (1990), and Wickens (1989) and to the comprehensive accounts by Agresti (1990), Bishop, Fienberg, and Holland (1975), and Haberman (1978, 1979).

Programs to perform the analyses discussed in this chapter are available in standard packages such as SPSS-X, BMDP, and SAS.

Odds and Odds Ratios

Loglinear analysis of a multivariate frequency table essentially amounts to carrying out analyses by means of odds and odds ratios. Tables 2.1 and 2.2 will be used to explain these basic concepts. Table 2.1 contains information on the distribution of education in the United States, and sheds light on the question of whether women have had the same educational opportunities as men.

From the marginal distribution of Education in Table 2.1 it is estimated that in the United States in 1981 the overall probability that someone from the population of 16 years and older has realized a high educational level ("Some College") equals .407, whereas the chances of a low educational level ("Less Than College") are .593. Consequently, the marginal odds of having attained a high rather than a low educational level are .407/.593 = 471/685 = .688. The inverse marginal odds "low education/high education" are 1/.688 = .593/.407 = 685/471 = 1.454. The marginal probability that someone has a low educational

TABLE 2.1
Sex (S) and Education (E)

S. Sex	*E. Education* 1. *Some College*	2. *Less Than College*	*Total*
1. Men	237 (.467)	271 (.533)	508 (1.000)
2. Women	234 (.361)	414 (.639)	648 (1.000)
Total	471 (.407)	685 (.593)	1156 (1.000)

SOURCE: The Political Action Study; see Barnes and Kaase (1979) and Jennings and Van Deth (1989).
NOTE: Proportions based on the row totals appear in parentheses.

level is almost 1.5 times higher than the probability of someone having a high educational level.

The conditional distributions of Education in Table 2.1 show that for men the odds of having attained a high rather than a low educational level are .875 (= .467/.533 = 237/271), which is slightly higher than the corresponding odds for women: .565 (= .361/.639 = 234/414). The way in which these two conditional odds differ from each other can be expressed by taking their ratio: .875/.565 = (237/271)/(234/414) = 1.547. Apparently, the odds of attaining a high rather than a low educational level are about 1.5 times as favorable for men as they are for women. Such a ratio of two conditional odds is called an *odds ratio*; it is a measure of association, in this case, between Sex and Education (Knoke & Burke, 1980, pp. 10-11).

An interesting question is whether this discrepancy in educational opportunities between men and women has increased over time or diminished from generation to generation. Table 2.2 contains the relevant data.

Table 2.1 can be regarded as a marginal table from Table 2.2, as Table 2.1 may be obtained by collapsing Table 2.2 over Age. As just shown, the marginal odds ratio in Table 2.1 measuring the differences between men and women with regard to their educational opportunities equals 1.547. The corresponding conditional odds ratios for each of the three age groups can be computed from the three Sex × Education subtables in Table 2.2. For the youngest age group, this conditional odds ratio is 1.337 [= (100/83)/(91/101)]; for the middle group, 1.745 [= (92/96)/(78/142)]; and for the oldest, 1.287 [= (45/92)/(65/171)]. The ways in

TABLE 2.2
Age (A), Sex (S), and Education (E)

A. Age	*S. Sex*	*E. Education* *1. Some College*	*2. Less Than College*	*Total*
1. 16-34	1. men	100	83	183
	2. women	91	101	192
	subtotal	191	184	375
2. 35-57	1. men	92	96	188
	2. women	78	142	220
	subtotal	170	238	408
3. 58-91	1. men	45	92	137
	2. women	65	171	236
	subtotal	110	263	373
	Total	471	685	1156

SOURCE: See Table 2.1.

which the three conditional odds ratios differ from each other can be expressed by taking their ratios. These ratios of odds ratios are called *second-order odds ratios* and are measures of the statistical three-variable interaction as they indicate to what extent the association between two variables varies among the categories of a third variable. Comparison of the conditional odds ratios of the young against the middle age group yields a second-order odds ratio of .766 (= 1.337/1.745); comparing the young to the old group gives 1.039 (= 1.337/1.287) and the middle to the old age group 1.356 (= 1.745/1.287).

As the conditional odds ratios show, in all three age groups the odds of attaining a high rather than a low educational level are somewhat less favorable for women than for men. The conditional odds ratios of the young and the old age groups are about the same and, accordingly, the pertinent second-order odds ratio is about 1. As follows from the other second-order odds ratios, the conditional odds ratio of the middle age group is 1.356 times higher than of the oldest group, and about the same difference exists between the middle and the youngest age groups (1/.766 = 1.305). The discrepancy between men and women's educational opportunities is the highest in the middle age group. So there is not a systematic linear trend over generations toward increasing or diminishing inequality of educational opportunities between men and women.

When interpreting and comparing the magnitudes of odds and odds ratios (and of the multiplicative parameters of the loglinear model that will be discussed in the next section), one should keep in mind that the values of odds and odds ratios are asymmetrically situated around 1, the value of "no difference"; the limit of the maximum "negative" difference is 0, and $+\infty$ is the limit for the maximum "positive" difference. Thus an odds ratio equal to 4.00 indicates the same amount of discrepancy between the two constituting odds as an odds ratio of .25 (= 1/4.00), albeit in a different direction. This asymmetry disappears when one works with the natural logarithm of odds and odds ratios: the "no difference" value equals $\ln 1 = 0$, the negative limiting value is $-\infty$, and the positive limit is $+\infty$. In this way, the sizes of the positive and negative values can be compared more easily: The "equal strength" odds 4 and .25 become $\ln 4 = +1.386$ and $\ln .25 = -1.386$.

The natural log-odds are usually termed *logits*. The conditional logit "high education/low education" for men in Table 2.1 equals $\ln (237/271) = \ln 237 - \ln 271 = -.134$ and the corresponding conditional logit for women equals $\ln 234 - \ln 414 = -.570$. The logarithm of the odds ratio in Table 2.1 is obtained by subtracting these two conditional logits from each other: $-.134 - (-.570) = .436$ (= ln 1.547). The logarithms of higher-order odds ratios may be found in an analogous manner by subtracting the logarithms of the appropriate lower-order odds ratios from each other.

These insights and basic concepts are essential to understanding the logic behind loglinear modeling. In addition, one should be familiar with the concepts of partial odds and partial odds ratios. Partial odds are average conditional odds, where the geometric mean is used as the measure of central tendency. The geometric mean is the nth root of the product of n numbers (Weisberg, 1992, pp. 41-43). As the common arithmetic mean is the logical choice as a measure of central tendency if one works with sums and differences (of probabilities)—that is, with an additive model—the geometric mean is the natural choice when working with products and divisions, with odds and odds ratios, that is, with a multiplicative model (see Hagenaars, 1990, pp. 32-33). For example, the geometric mean of the numbers 10 and 1,000, equal to $\sqrt{(10 \times 1000)} = 100$, is the appropriate central value in the sense that the number 10 is ten times lower and the number 1,000 is ten times higher than 100.

The partial odds "high education/low education" for Table 2.1 are .703 $[=\sqrt{(.875)(.565)}]$, the geometric mean of the two conditional odds "high/

low education" for men and women. This value is not identical to the value .688 obtained for the corresponding marginal odds in Table 2.1. In general, partial odds and partial odds ratios will not have the same values as their marginal counterparts.

The partial odds ratio is defined as the geometric mean of the corresponding conditional odds ratios. Three conditional odds ratios indicating the association between Sex and Education for three age groups were discussed above (Table 2.2). The partial odds ratio "Sex by Education" in Table 2.2 is the geometric mean of these three conditional odds ratios and equals $\sqrt[3]{(1.337)(1.745)(1.287)} = 1.443$. This is slightly less than the corresponding marginal odds ratio of 1.547 obtained from Table 2.1. While the marginal odds ratio measures the overall association between Sex and Education (in Table 2.1), the partial odds ratio is a measure of the partial association between Sex and Education, holding Age constant (in Table 2.2).

Besides being the average of the three conditional odds ratios Sex-Education, the partial odds ratio Sex-Education may also be conceived of as the ratio of the partial odds "high education/low education" for men and women. In Table 2.2 there are three conditional odds "high education/low education" for men, one for each age group. Taking the geometric mean of these three conditional odds gives the partial odds "high education/low education" for men yielding the value .826. The corresponding partial odds for women are .573. The ratio of these two partial odds is mathematically equivalent to the geometric mean of the three conditional odds ratios discussed above and gives (within rounding errors) the same result: .826/.573 = 1.442.

How well the value of the partial odds ratio approximates the values of the corresponding conditional odds ratios is indicated by the second-order odds ratios. The latter indicate how much the conditional odds ratios differ from each other, and thus from the partial odds ratio.

Saturated Loglinear Models

The informal analysis of Table 2.2 above can be carried out more formally and completely by means of loglinear modeling. The following notation will be used. Observed frequencies are indicated by f; superscripts refer to variables and subscripts to categories of these variables. For example, the first observed cell frequency in Table 2.2 is indicated by f_{111}^{ASE} (= 100). The total sample size is denoted by N. The observed proportions p are defined as $p = f/N$. Probabilities in the population are denoted by π and expected cell frequencies F are defined as $F = N\pi$. F^{ASE}_{ijk} then represents the number of people who would have been found in cell i,j,k of table ASE if the sample were an exact reflection of the population without sampling fluctuations. Maximum likelihood estimates of F and π are denoted by $\hat{F}$ and $\hat{\pi}$. The + sign as subscript means that the frequencies or probabilities are summed over the replaced subscript. For example, the cell frequencies in Table 2.1 may be obtained by summing the frequencies of Table 2.2 over Age, indicated by f^{ASE}_{+jk}.

Denoting the parameters of the loglinear model by η and τ, the general, nonrestrictive loglinear model for table ASE in its multiplicative form is

$$F^{ASE}_{ijk} = \eta\tau^{A}_{i}\tau^{S}_{j}\tau^{E}_{k}\tau^{AS}_{ij}\tau^{AE}_{ik}\tau^{SE}_{jk}\tau^{ASE}_{ijk}, \qquad (1)$$

where $i = 1, \ldots, I$; $j = 1, \ldots, J$; $k = 1, \ldots, K$.

The additive form, which gives the model its name, is

$$G^{ASE}_{ijk} = \theta + \lambda^{A}_{i} + \lambda^{S}_{j} + \lambda^{E}_{k} + \lambda^{AS}_{ij} + \lambda^{AE}_{ik} + \lambda^{SE}_{jk} + \lambda^{ASE}_{ijk}, \qquad (2)$$

where $G = \ln F$, $\theta = \ln \eta$, $\lambda = \ln \tau$.

This is not an identifiable model; there are more unknown parameters to be estimated than known cell frequencies F. Identifying restrictions on the parameters are necessary. One possibility is to express all effects in terms of deviations from the effects in a particular category of each variable and to set all τ-parameters referring to that particular category equal to the "no effect" value of 1 and the corresponding λ-parameters to 0 (e.g., $\tau^{ASE}_{1jk} = \tau^{ASE}_{i1k} = \tau^{ASE}_{ij1} = 1$; $\lambda^{ASE}_{1jk} = \lambda^{ASE}_{i1k} = \lambda^{ASE}_{ij1} = 0$). This corresponds with "dummy coding" of variables, often used in regression analysis to include nominal level variables in the regression equation (Lewis-Beck, 1980, pp. 66-71).

An alternative way of defining identifying restrictions—and the one that will be used exclusively throughout this monograph—is to express each effect in terms of deviations from the average effect and to impose the restriction that each τ-parameter multiplied over any of its subscripts equals 1 and the corresponding λ-parameter summed over any of its subscripts equals 0 (e.g., $\prod_i \tau^{ASE}_{ijk} = \prod_j \tau^{ASE}_{ijk} = \prod_k \tau^{ASE}_{ijk} = 1$; $\Sigma_i \lambda^{ASE}_{ijk} = \Sigma_j \lambda^{ASE}_{ijk} = \Sigma_k \lambda^{ASE}_{ijk} = 0$). Instead of dummy coding, now effect coding is used for the variables, a common practice in analysis of variance (Hays, 1981, pp. 331-333).

Dummy coding and effect coding generally yield very different parameter estimates. However, when interpreted correctly taking the different coding schemes into account, exactly the same substantive conclusions are drawn. Readers not familiar with these different parametrizations of "effects" are urged to study the articles by Alba (1987, also in Long, 1988), Kaufman and Schervish (1986, 1987), and Long (1984).

The loglinear model in Equations 1 and 2 is called a *saturated* model. With the identifying restrictions imposed on the parameters, an exactly identified model results that imposes no restrictions on the data. For the saturated model, the observed frequencies f are the maximum likelihood estimates of F, so $\hat{F} = f$ (assuming the data follow a [product]multinomial sampling distribution). By substituting f for F in Equation 1, the maximum likelihood estimates $\hat{\tau}$ can be computed. The relevant formulas are presented by, among others, Knoke and Burke (1980) and Hagenaars (1990, p. 38).

Table 2.3 shows the parameter estimates for the saturated model applied to Table 2.2. Effect coding was used and the interpretation of the parameter estimates should take this into account.

The parameters of the loglinear model are directly related to odds and odds ratios. Because the purpose of this monograph is to introduce latent variable models, this relationship is explained only in a concise and rather intuitive manner. For those readers having difficulties with the expositions below, again, the articles by Alba (1987, also in Long, 1988), Kaufman and Schervish (1986, 1987), and Long (1984) will be most helpful.

The overall effect η is not a very interesting parameter. It tells us that the geometric mean of all cell frequencies in Table 2.2 equals 91.5, a mere reflection of the sample size.

One-variable effects are directly related to partial odds. The value of $\hat{\tau}_1^A$ (1.022) indicates that the geometric mean of the four frequencies $\hat{F}_{1jk}^{ASE}$ pertainingto $A = 1$ (100, 83, 91, 101) is 1.022 times bigger than expected from the overall effect η. From the value of $\hat{\tau}_2^A$, it follows that the geometric mean of the four frequencies $\hat{F}_{2jk}^{ASE}$ is 1.087 times bigger than expected on the basis of the overall effect. In other words, the partial odds of being 16-34 years old and not 37-57 are 1.022/1.087 = .940. Following the same line of reasoning, it can be concluded that the partial odds of being a man and not a woman are .896/1.116 = $.896^2$ = .803 and the partial odds of having realized a high rather than a low educational level is .830/1.205 = $.830^2$ = .689.

TABLE 2.3

The Saturated Model for Table 2.2

Effect of		$\hat{\tau}$	$\hat{\lambda}$	$\hat{s}_{\lambda}$[a]	$z = \hat{\lambda}/\hat{s}_{\lambda}$
Overall		91.478	4.516		
A (Age)	1 = 16-34	1.022	.021	.043	.497
	2 = 35-57	1.087	.084	.043	1.960
	3 = 58-91	.900	–.105	.046	–2.295
S (Sex)	1 = men	.896	–.110	.031	–3.537
	2 = women	1.116	.110	.031	3.537
E (Education)	1 = some college	.830	–.187	.031	–6.026
	2 = less than college	1.205	.187	.031	6.026
AS	11	1.088	.084	.043	1.954
	12	.919	–.084	.043	–1.954
	21	1.054	.053	.043	1.244
	22	.949	–.053	.043	–1.244
	31	.872	–.137	.046	–3.000
	32	1.147	.137	.046	3.000
AE	11	1.230	.207	.043	4.813
	12	.813	–.207	.043	–4.813
	21	1.027	.026	.043	.619
	22	.974	–.026	.043	–.619
	31	.792	–.234	.046	–5.110
	32	1.263	.234	.046	5.110
SE	11	1.096	.092	.031	2.955
	12	.912	–.092	.031	–2.955
	21	.912	–.092	.031	–2.955
	22	1.096	.092	.031	2.955
ASE	111	.981	–.019	.043	–.440
	112	1.019	.019	.043	.440
	121	1.019	.019	.043	.440
	122	.981	–.019	.043	–.440
	211	1.048	.048	.043	1.115
	212	.954	–.048	.043	–1.115
	221	.954	–.048	.043	–1.115
	222	1.048	.048	.043	1.115
	311	.972	–.029	.046	–.625
	312	1.029	.029	.046	.625
	321	1.029	.029	.046	.625
	322	.972	–.029	.046	–.625

a. $\hat{s}_{\lambda}$ denotes the estimated standard error of $\hat{\lambda}$; see text.

Usually more interesting from a substantive point of view are the two-variable parameters. They are directly related to partial odds ratios.

For example, the $\hat{\tau}_{jk}^{SE}$ parameters show that on average, within the categories of Age, the cell frequencies referring to the categories (1,1) and (2,2) of the joint variable *SE* are 1.096 times larger than expected on the basis of the lower-order effects, whereas the cells (1,2) and (2,1) are .912 times smaller. From this (and more formally from the formulas for computing the values of the two-variable parameters, not presented here) it follows that the partial odds ratio of 1.443 for Sex and Education discussed in the previous section can be expressed (within rounding errors) in terms of the $\hat{\tau}_{jk}^{SE}$ parameters: $(\hat{\tau}_{11}^{SE}/\hat{\tau}_{12}^{SE})/(\hat{\tau}_{21}^{SE}/\hat{\tau}_{22}^{SE}) = (1.096/.912)/(.912/1.096) = 1.096^4 = 1.444$. We can therefore conclude that, controlling for Age, the partial odds of having had a high and not a low education are 1.44 times more favorable for men than for women.

Applying the same routines to the other two-variable parameters, it can be concluded that, holding Sex constant, educational opportunities have systematically increased from one generation (age category) to another and that, on average, within categories of Education, there are more women in the oldest age category than men, relative to the other age categories.

The three-variable parameters represent statistical three-variable interaction effects and are closely related to second-order odds ratios. For example, it can be deduced from the estimated three-variable parameters that the conditional odds ratio Sex-Education for category 1 (young) of Age is .927 $[= (\hat{\tau}_{111}^{ASE}/\hat{\tau}_{112}^{ASE})/(\hat{\tau}_{121}^{ASE}/\hat{\tau}_{122}^{ASE}) = (.981/1.019)/(1.019/.981)]$ times smaller than the partial odds ratio Sex-Education, while the corresponding conditional odds ratio for category 2 (middle) of Age is 1.207 $[= (\hat{\tau}_{211}^{ASE}/\hat{\tau}_{212}^{ASE})/(\hat{\tau}_{221}^{ASE}/\hat{\tau}_{222}^{ASE}) = (1.048/.954)/(.954/1.048)]$ times larger than this partial odds ratio. So the conditional odds ratio Sex-Education for the young age group is .768 (= .927/1.207) times smaller than for the middle age group, which is, within rounding errors, the same result as obtained in the previous section.

All three-variable effects in Table 2.3 are very small. They are also not statistically significant, as follows from the information contained in the last two columns of Table 2.3. The second-to-last column provides the estimated standard errors of the $\hat{\lambda}$-parameters (Agresti, 1990, chap. 12; Knoke & Burke, 1980, p. 19). The last column presents the test statistic z, which approximates the standard normal distribution if the null hypothesis of no effect $\lambda = 0$ is true. The outcomes of the z tests raise the question whether the three-variable effects are really needed for an adequate description of the relationships among the variables in Table 2.2, a question to be answered in the next section.

Nonsaturated Loglinear Models

In contrast to saturated models, nonsaturated models impose a priori restrictions on the data. For example, we might assume that Age, Sex, and Education have direct relations with each other, but that there are no three-variable interactions. The relation between Sex and Education is the same in all three Age categories. In loglinear terms, this hypothesis implies a model in which all three-variable effects in Equation 1 have been set to one.

An important class of nonsaturated loglinear models is the class of hierarchical models. A model is said to be hierarchical if the presence of a higher-order parameter implies the presence of all lower-order parameters that can be formed from the superscripts of the higher-order parameter. For example, if parameter τ_{ij}^{AS} is not a priori fixed at one, then τ_i^A and τ_j^S are also not a priori set to one. Hierarchical models can thus be represented by their highest-order terms. Model $\{ASE\}$ denotes the (hierarchical) saturated model for Table 2.2, and model $\{AS, AE, SE\}$ the above-mentioned no three-variable interaction model.

Hierarchical models have the property that the observed marginal tables corresponding with the highest-order terms of the model are exactly reproduced by the estimated expected frequencies. For example, the estimated expected frequencies for model $\{AS, AE, SE\}$ are such that $\hat{F}_{ij+}^{ASE} = f_{ij+}^{ASE}$, $\hat{F}_{i+k}^{ASE} = f_{i+k}^{ASE}$, $\hat{F}_{+jk}^{ASE} = f_{+jk}^{ASE}$.

Finding the maximum likelihood estimates $\hat{F}$ for a particular nonsaturated model may require an iterative solution. Most often, the iterative proportional fitting algorithm or the Newton/Raphson procedure is used. For further details on these procedures, the interested reader is referred to the works cited at the beginning of this chapter.

Once the model has been estimated, it is possible to test the empirical validity of the nonsaturated model by comparing the observed frequencies with the estimated expected frequencies using the Pearson chi-square χ^2 or the log-likelihood ratio chi-square statistic L^2 (Knoke & Burke, 1980, p. 30).

The number of degrees of freedom equals the number of independent a priori restrictions that have been imposed on the model parameters. The no three-variable interaction model $\{AS, AE, SE\}$ for Table 2.2 implies a priori restrictions on the three-variable parameters τ_{ijk}^{ASE}. There are $3 \times 2 \times 2 = 12$ parameters τ_{ijk}^{ASE}. Because of the identifying restrictions, the number of independent τ_{ijk}^{ASE}-parameters equals $(3-1)(2-1)(2-1) = 2$. So, model $\{AS, AE, SE\}$ has two degrees of freedom. The test results for this model are $L^2 = 1.247$, df = 2, $p = .54$ ($\chi^2 = 1.245$). There is no reason to reject the hypothesis that there are no three-variable effects in the population. The somewhat higher observed discrepancy between the educational opportunities of men and women in the middle age group is probably caused by sampling fluctuations.

The parameter estimates for the one-variable effects *A, S,* and *E* and the two-variable effects *AS, AE,* and *SE* in this restricted model are almost identical to the estimates in Table 2.3, which is not surprising in light of the insignificant and very small values of the three-variable effects in the saturated model. (The estimates are reported in Table 5.3, in Chapter 5.)

More restrictive models for the data in Table 2.2 have to be rejected. For example, model $\{AS, AE\}$, in which there is no relation at all between Sex and Education, yields the following results $L^2 = 10.724$, df = 3, $p = .01$ ($\chi^2 = 10.724$). In a way this is a borderline result: Employing a .05 confidence level, the model has to be rejected; with a .01 level, it does not have to be rejected. However, the more powerful conditional test for hierarchically nested models (Agresti, 1990, sec. 4.4.2; Hagenaars, 1990, sec. 2.5) in which the validity of the *r*estricted model $\{AS,AE\}$ is tested, assuming the *u*nrestrictive (but nonsaturated) model $\{AS,AE,SE\}$ to be valid in the population, yields $L^2_{r/u} = L^2_r - L^2_u = 10.724 - 1.247 = 9.478$, $df_{r/u} = df_r - df_u = 3 - 2 = 1$, $p = .002$. Model $\{AS,AE\}$ must be rejected in favor of the less restrictive model $\{AS,AE,SE\}$ with the direct relation between Sex and Education.

Logit or Effect Models

The loglinear models dealt with so far have involved symmetrical relationships among the variables in which no distinction was made in terms of dependent and independent variables. However, the relations among the variables in Table 2.2 are asymmetrical: Age and Sex may be regarded as causes of Education, but Education cannot be seen as a cause of Age or Sex. This implicit asymmetrical order of the variables can be made explicit by focusing on the effects of the independent variables Age and Sex on the dependent variable Education and by defining as the "dependent quantity" of the loglinear equations not the cell frequencies F^{ASE}_{ijk} but the conditional odds of having obtained a high instead of a low educational level.

The conditional odds are denoted by Ω:

$$\Omega_{ij}^{AS\bar{E}} = \pi_{ij1}^{ASE}/\pi_{ij2}^{ASE} = F_{ij1}^{ASE}/F_{ij2}^{ASE}, \qquad (3)$$

where a bar is placed above the dependent variable. Sometimes the dependent variable is provided with a subscript k/k' ($k \neq k'$) to indicate that the odds refer to the probabilities of belonging to category k instead of k': $\Omega_{ij\,k/k'}^{AS\bar{E}}$. If the dependent variable is dichotomous, then $k = 1$ and $k' = 2$ and the subscripts are usually omitted (as was done in Equation 3). $\hat{\Omega}$ is the maximum likelihood estimator of Ω and is defined analogously in terms of the estimated probabilities $\hat{\pi}$ or the estimated expected frequencies $\hat{F}$. The corresponding log-odds or logits are indicated by Φ, defined as

$$\Phi_{ij}^{AS\bar{E}} = \ln \Omega_{ij}^{AS\bar{E}} = \ln \pi_{ij1}^{ASE} - \ln \pi_{ij2}^{ASE} = \ln F_{ij1}^{ASE} - \ln F_{ij2}^{ASE}. \qquad (4)$$

Denoting the parameters of the multiplicative effect model by γ and the parameters of the logit model by β, the saturated effect or logit model for Table 2.2 may be written as

$$\Omega_{ij}^{AS\bar{E}} = \gamma^{\bar{E}} \gamma_i^{A\bar{E}} \gamma_j^{S\bar{E}} \gamma_{ij}^{AS\bar{E}}$$

$$\Phi_{ij}^{AS\bar{E}} = \beta^{\bar{E}} + \beta_i^{A\bar{E}} + \beta_j^{S\bar{E}} + \beta_{ij}^{AS\bar{E}}, \qquad (5)$$

where, as before, effect coding is applied: All γ-parameters multiplied over any subscript equal 1 and the β-parameters summed over any subscript equal 0.

Nonsaturated models can be obtained by deleting particular effects. Testing and fitting a logit model, estimating its parameters, and interpreting them may be accomplished most easily by linking effect models to regular loglinear models. Actually, the effect models discussed here are just ordinary loglinear models, and all that has been said about the latter applies to the former as well (Knoke & Burke, 1980, pp. 24-29). The following equations may clarify this:

$$\Omega_{ij\,1/2}^{AS\bar{E}} = \frac{F_{ij1}^{ASE}}{F_{ij2}^{ASE}} = \frac{\eta \tau_i^A \tau_j^S \tau_1^E \tau_{ij}^{AS} \tau_{i1}^{AE} \tau_{j1}^{SE} \tau_{ij1}^{ASE}}{\eta \tau_i^A \tau_j^S \tau_2^E \tau_{ij}^{AS} \tau_{i2}^{AE} \tau_{j2}^{SE} \tau_{ij2}^{ASE}} \qquad (6)$$

$$= \frac{\tau_1^E}{\tau_2^E} \cdot \frac{\tau_{i1}^{AE}}{\tau_{i2}^{AE}} \cdot \frac{\tau_{j1}^{SE}}{\tau_{j2}^{SE}} \cdot \frac{\tau_{ij1}^{ASE}}{\tau_{ij2}^{ASE}}$$

$$= \gamma^{\bar{E}}_{1/2} \cdot \gamma^{A\bar{E}}_{i\,1/2} \cdot \gamma^{S\bar{E}}_{j\,1/2} \cdot \gamma^{AS\bar{E}}_{ij\,1/2} \cdot$$

A particular effect model is equivalent to that ordinary loglinear model in which all corresponding terms linking the dependent with independent variables occur, *plus* all parameters pertaining to the relationships among the independent variables. In effect models, the joint observed frequency distribution of the independent variable(s) is always exactly reproduced by the estimated expected frequencies $\hat{F}$.

Because of this correspondence of logit and regular loglinear models, polytomous dependent variables pose no special problem. If the dependent variable is trichotomous with categories 1, 2, 3, the effects on the odds ½ and the odds ⅓ can be found by taking the ratio of the relevant τ-parameters of the appropriate loglinear model, analogous to Equation 6.

Modified Path Models

Goodman (1972) coined a very appropriate name for effect or logit models: *modified multiple regression approach.* This label is aptly chosen because there is a striking analogy between ordinary regression models and logit models. Both are aimed at estimating the direct effects of several independent variables on one dependent variable, and both take the relations among the independent variables themselves for granted.

When the relationships among the independent variables are also of interest, path or structural models consisting of a system of several regression equations have to be set up. In an analogous manner, systems of logit equations can be defined. When introducing these systems of logit equations, Goodman (1973) called them "a modified path analysis approach." The approach can be illustrated using the simple three-variable models depicted in Figure 2.1.

If the variables in Figure 2.1a are measured at interval level, the causal path diagram 2.1a is equivalent to the following system of regression equations (Asher, 1976):

$$X_2 = a_2 + b_{21}X_1 + e_2 \qquad (7)$$

$$X_3 = a_3 + b_{31}X_1 + b_{32}X_2 + e_3 \qquad (8)$$

When determining the magnitude of a particular direct effect in a certain path model, the variables that occur later in the causal sequence are

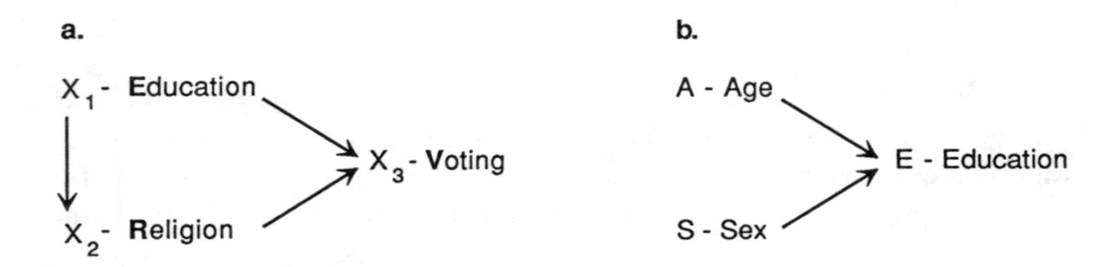

Figure 2.1. Causal Models

ignored. The effect of Education (X_1) on Religion (X_2) is determined in Equation 7 by the zero-order regression coefficient b_{21} and not by the partial coefficient $b_{21.3}$. Because Voting (X_3) is causally posterior to Education and Religion, it cannot in reality influence the relationship between Education and Religion. Holding Voting constant when estimating the relation between Education and Religion would be nonsensical. It would also be misleading, as the zero-order coefficient b_{ji} is usually different from the partial coefficient $b_{ji.k}$.

As Voting is directly influenced by more than one variable, the coefficients b_{31} and b_{32} in Equation 8, although not written as such, are partial coefficients, namely, $b_{31.2}$ and $b_{32.1}$, respectively. If causal model 2.1a had been extended with one or more variables, all of them causally posterior to Voting, Equation 8 would not change and these extra variables would not be held constant. Furthermore, in standard path or recursive models (Asher, 1976, chap. 3) the separate equations are ordinary multiple regression equations, in which, for estimating the coefficients of the equation, the relations among the independent variables occurring in that equation are taken for granted.

Treating the variables in model 2.1a as categorical, model 2.1a represents a modified path model consisting of a system of two ordinary logit equations. The effect of Education (now symbolized by E instead of X_1) on Religion (R) is determined using the logit equation

$$\Omega_i^{E\bar{R}} = \gamma^{\bar{R}}\,\gamma_i^{E\bar{R}}\,. \tag{9}$$

The logit model in Equation 9 corresponds with loglinear model $\{ER\}$ applied to table ER. The second logit equation, for determining the effects of Education and Religion on Voting, is

$$\Omega_{i\,j}^{ER\bar{V}} = \gamma^{\bar{V}}\, \gamma_i^{E\bar{V}}\, \gamma_j^{R\bar{V}} \,. \tag{10}$$

Now the corresponding loglinear model is model $\{ER,EV,RV\}$ for table ERV. As in path analysis, because of the presumed causal order of the variables, the effect of E on R is determined in table ER without holding V constant. Moreover, in Equation 10 the relation between E and R is taken for granted.

There is thus a rather close analogy between path analysis and the modified path analysis approach (see also Kiiveri & Speed, 1982; Whittaker, 1990). But there are also several differences. There is no "calculus of path coefficients" for loglinear path models (Fienberg, 1980, chap. 7). Although it is possible to compute indirect and total effects (see Chapter 5), there are no rules for decomposing the total effect in terms of direct and indirect effects.

Another difference is that the model in Figure 2.1a is a saturated model when viewed as a path diagram because it reproduces exactly the observed covariance matrix. As is clear from Equation 10, the corresponding modified path model is not saturated because it excludes the interaction effects of Education and Religion on Voting. (In the graphic representations used in this monograph, effects of three or more variables will be denoted by knots connecting the three or more variables; see Figure 4.1, Chapter 4). Goodman (1973) has shown how to test nonsaturated modified path models. This is illustrated using the data in Table 2.2 and the causal diagram in Figure 2.1b.

According to model 2.1b, it is first hypothesized that Age and Sex influence Education directly without affecting each other's influence on Education. The second hypothesis is that Age and Sex are independent of each other. The latter hypothesis may be of interest when one is investigating differential mortality of men and women, or, more likely in the context of these data, when the investigator believes that Age and Sex are unrelated to each other in the population, but wants to test whether differential nonresponse of men and women for several age groups might have distorted the data.

Loglinear model $\{AE,SE\}$ is not an adequate representation of the path diagram in Figure 2.1b and its two implied hypotheses. An implication of model $\{AE, SE\}$ is that Age and Sex are unrelated to each other, holding Education constant. But given the assumed causal order in path diagram 2.1b, whether there is statistical independence between Age and Sex has to be tested without holding Education constant, that is, using marginal table AS.

A system of two logit equations and a stepwise testing procedure is needed. First, marginal table *AS* is set up with frequencies f^{ASE}_{ij+}. The independence model $\{A, S\}$ for marginal table *AS* yields the following test results: $L^2 = 12.325$, df = 2, $p = .002$ ($\chi^2 = 12.226$).

Second, the hypothesis about the effects of Age and Sex on Education is tested by means of the logit model that corresponds with loglinear model $\{AS, AE, SE\}$ for table *ASE*. The test results are $L^2 = 1.247$, df = 2, $p = .536$ ($\chi^2 = 1.245$) (see also the section above on nonsaturated loglinear models).

The overall test of the compound hypothesis H^* that the whole path diagram in Figure 2.1b is true—that is, that simultaneously submodel $\{A, S\}$ is valid for marginal table *AS* and submodel $\{AS, AE, SE\}$ for the full table *ASE*—can be obtained by summing the separate L^2 values of the submodels and their degrees of freedom. Denoting the overall log-likelihood ratio chi-square test statistic by L^*, the results are $L^* = 12.325 + 1.247 = 13.572$, $\text{df}^* = 2 + 2 = 4, p < .01$. The hypotheses implied by the path diagram have to be rejected. As the separate tests for the submodels show, the reason for this rejection is the fact that model $\{A, S\}$ does not fit the collapsed table *AS*.

Introducing the "extra" effect *AS* into the path diagram implies fitting the saturated model $\{AS\}$ to the collapsed table *AS*, which by definition fits the data perfectly. Testing the new path model as a whole now amounts to testing the validity of loglinear model $\{AS, AE, SE\}$ for the complete table *ASE*. Nevertheless, the collapsed table *AS* is still needed to obtain the wanted estimated association between Age and Sex. The relevant parameter estimates in the saturated model $\{AS\}$ for table *AS* are $\hat{\lambda}^{AS}_{11} = .101$, $\hat{\lambda}^{AS}_{21} = .046$, and $\hat{\lambda}^{AS}_{31} = -.147$, which indeed differ slightly from the corresponding effects in model $\{AS, AE, SE\}$ applied to the complete table *ASE*; these are .082, .045, −.127, respectively.

Sometimes marginal and partial effects are not different from each other. There is a theorem, the collapsibility theorem, that indicates under which conditions partial and marginal loglinear parameters are identical (Bishop et al., 1975, p. 47; Hagenaars, 1990, sec. 2.6).

Goodman (1972) also shows how to compute the estimated expected frequencies $\hat{F}^*$ assuming the compound hypothesis H^* is true:

$$\hat{F}^{*ASE}_{ijk} = \hat{F}^{AS}_{ij} \frac{\hat{F}^{ASE}_{ijk}}{f^{AS}_{ij}}, \tag{11}$$

where $\hat{F}^{AS}_{ij}$ are the estimated expected frequencies of model $\{A,S\}$ for table AS and $\hat{F}^{ASE}_{ijk}$ are the estimated expected frequencies of model $\{AS, AE, SE\}$ for table ASE. The frequencies $\hat{F}^{*}$ can be used to calculate L^{*} directly; furthermore, the estimated effect parameters for the whole model can be obtained using the frequencies $\hat{F}^{*}$, where necessary, given the causal order of the variables, summed over the appropriate subscripts.

It may be illuminating to render Equation 11 in terms of probabilities. Let the $\hat{\pi}$'s in Equation 12 have analogous meanings to the $\hat{F}$'s in Equation 8. We may then write:

$$\hat{\pi}^{*ASE}_{ijk} = \hat{\pi}^{AS}_{ij} \frac{\hat{\pi}^{ASE}_{ijk}}{p^{AS}_{ij}} . \qquad (12)$$

Remembering that the estimated expected frequencies of the separate logit equations are such that for each logit model the observed (joint) frequency distribution of the independent variables is exactly reproduced, Equation 12 can be rewritten as

$$\hat{\pi}^{*ASE}_{ijk} = \hat{\pi}^{AS}_{ij} \frac{\hat{\pi}^{ASE}_{ijk}}{\hat{\pi}^{ASE}_{ij+}} = \hat{\pi}^{AS}_{ij} \hat{\pi}^{AS\bar{E}}_{ijk} , \qquad (13)$$

where $\pi^{AS\bar{E}}_{ijk}$ is the estimated conditional probability of belonging to category k of E, given one belongs to category (i,j) of the joint variable AS.

Although the verbal description becomes cumbersome, the essential stepwise approach of modified path modeling embodied in these equations may be formulated as follows: The estimated probability that a randomly selected individual obtains scoring pattern (i,j,k) on the joint variable ASE if H^{*} is valid in the population is equal to the estimated probability $\hat{\pi}^{AS}_{ij}$ that, given that model $\{A,S\}$ is true, this individual belongs to $A = i$ and $S = j$; multiplied by the estimated conditional probability $\hat{\pi}^{AS\bar{E}}_{ijk}$ that, model $\{AS, AE, SE\}$ being true, this person belongs to $E = k$, given $AS = (i,j)$.

Only the bare essentials of loglinear modeling have been presented here. Because knowledge of these essentials is prerequisite to understanding the remainder of this book, readers having difficulties with these basic concepts should consult the references mentioned.

3. THE LATENT CLASS MODEL

Like many brilliant ideas in social science methodology, the basic notions of latent class analysis were developed by Paul Lazarsfeld and his associates during the early 1950s (Lazarsfeld, 1950a, 1950b; Lazarsfeld & Henry, 1968). Credit for feasible and flexible algorithms for testing the validity of a wide variety of latent class models and estimating their parameters is due especially to Goodman (1974a, 1974b) and Haberman (1979). Introductions into these more recent developments are provided by Clogg (1981b), Formann (1985), Hagenaars (1990), Langeheine and Rost (1988), and McCutcheon (1987).

Latent class analysis programs are not yet part of standard packages such as SPSS-X, BMDP, and SAS. However, stand-alone programs are widely available for both PCs and mainframes (see Appendix A).

The Basic Model

There are two kinds of variables in latent class models. First, there are the directly observed manifest variables. These act as indicators for the second kind, the not directly observed latent variables. In the standard latent class model, both manifest and latent variables are treated as categorical nominal-level variables. Table 3.1 will be used to explain the basic principles of latent class analysis. All variables in this example, latent and manifest, are dichotomous; an example of the more general polytomous case is provided by Hagenaars and Halman (1989).

The data in Table 3.1 are taken from the Political Action Study (Barnes & Kaase, 1979; Jennings & Van Deth, 1989). In this study, many questions were asked about the forms and backgrounds of the respondents' political actions. The data in Table 3.1 pertain to the most basic concepts and indices that were developed in the course of this study. Each index in Table 3.1 can be conceived of as a typology of an important aspect of the respondents' political orientations. The question is whether it is possible to reduce these five indices to an even more basic and general orientation toward politics.

It is assumed that there is a latent variable X that represents the fundamental political orientation of the respondents. For the time being, this latent variable is regarded as a categorical variable with two (latent) classes, liberal or left-wing versus conservative or right-wing political orientation. People's scores on the five observed indices are a direct result of their belonging to one of these latent classes. However, the

TABLE 3.1
Indices of Political Orientation

A. System Responsive- ness	*B. Ideological Level*	*C. Repression Potential*	*D. Protest Approval*	*E. Conventional Participation* *1. Low*	*2. High*	*Total*
1. Low	1. (near) Ideologues	1. Low	1. Low	3	3	6
			2. High	10	26	36
		2. High	1. Low	4	19	23
			2. High	7	32	39
	2. Nonideologues	1. Low	1. Low	28	18	46
			2. High	48	54	102
		2. High	1. Low	109	68	177
			2. High	59	44	103
2. High	1. (near) Ideologues	1. Low	1. Low	3	12	15
			2. High	8	55	63
		2. High	1. Low	7	38	45
			2. High	10	63	73
	2. Nonideologues	1. Low	1. Low	16	16	32
			2. High	33	80	113
		2. High	1. Low	49	92	141
			2. High	46	96	142
			Total	440	716	1156

SOURCE: See Table 2.1.
NOTE: Variable *A*, System Responsiveness, is an index based on agreement with three statements such as "I don't think that public officials care much about what people like me think." Agreement with these items points to a low opinion of the system's responsiveness (1, low); disagreement points to a high opinion (2, high). Variable *B*, Ideological Level, indicates whether the respondent used left and right ideological concepts when describing political parties (1, [near] ideologues) or whether he or she had an idiosyncratic or false understanding of the nature of political parties (2, nonideologues). Variable *C*, Repression Potential, is an index based on approval of four actions, such as "the police using force against demonstrators"; disapproval of these actions points to a low repression potential (1, low) and approval to a high potential (2, high). Variable *D*, Protest Approval, is based on approval of seven protest actions ranging from "signing a petition" to "occupying buildings or factories" and "blocking traffic"; disapproval of these actions points to a low tolerance level (1, low) and approval to a high level (2, high). Variable *E*, Conventional Participation, is an index based on how often the respondent engages in conventional political activities ranging from "reading about politics in the newspapers" to "attending a political meeting or rally"; not taking part in these activities indicates a low level of political participation (1, low) and frequent participation points to a high level (2, high). The indices were dichotomized as closely as possible to the median value; missing data for variables *A*, *D*, and *E* were assigned to category 1; for variable *C*, to category 2.

relation between the latent variable and the manifest indicators is not deterministic, but probabilistic. Because of all kinds of "outside influences" (including measurement errors), people who belong to the liberal latent class will not always give liberal manifest answers, albeit with a greater probability than people from the conservative latent class, and

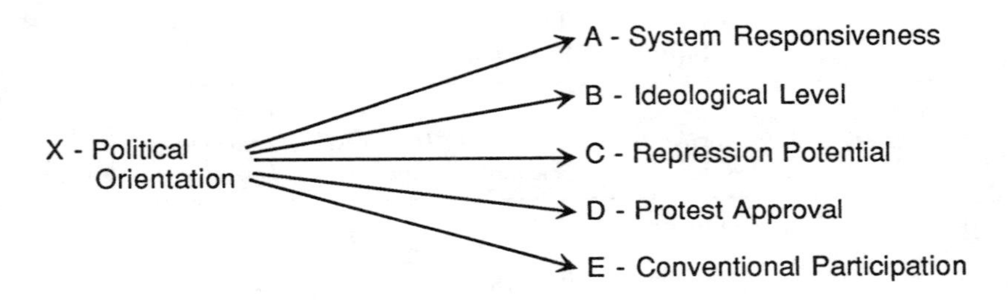

Figure 3.1. Basic Latent Class Model for Table 3.1
NOTE: Manifest variables, *A, B, C, D, E*; latent variable, *X*.

people from the conservative latent class sometimes choose the liberal alternative on a manifest item, although their chances of doing this are smaller than for the true liberals.

Furthermore, it is assumed that the scores on the manifest variables do not influence each other directly and that the outside influences and measurement errors affecting the score on a particular manifest item are independent of the outside influences affecting other manifest items. All the items have in common is their being indicators of the same latent variable. The manifest variables are correlated with each other, but this correlation disappears when the latent variable is held constant. This is the assumption of local independence, an assumption fundamental to all latent structure models, including factor analysis and latent class models. Figure 3.1 presents the path diagram for this latent class model.

The latent class model may be represented in terms of loglinear modeling or in terms of Lazarsfeld's original parameterization using conditional response probabilities. Given the nature of this monograph, most attention will be paid to the loglinear parameterization. Nevertheless, the perhaps more familiar Lazarsfeld parameterization, which is described in detail by McCutcheon (1987), will be presented first.

Following Goodman's (1974a, 1974b) notation, the basic equations of the latent class model in the case of five manifest variables A through E and one latent variable X with T categories may be stated as follows:

$$\pi_{i\,j\,k\,l\,m}^{ABCDE} = \sum_{t=1}^{T} \pi_{i\,j\,k\,l\,m\,t}^{ABCDEX}, \qquad (14)$$

where

$$\pi_{ijklmt}^{ABCDEX} = \pi_t^X \, \pi_{ijklmt}^{\overline{ABCDEX}} = \pi_t^X \, \pi_{it}^{\bar{A}X} \, \pi_{jt}^{\bar{B}X} \, \pi_{kt}^{\bar{C}X} \, \pi_{lt}^{\bar{D}X} \, \pi_{mt}^{\bar{E}X} \, . \tag{15}$$

The notation is essentially the same as in Chapter 2, with X being a latent variable with not directly observed scores. π_t^X is the probability that a person belongs to $X = t$ and as such is a measure of the size of latent class t. $\pi_{ijklmt}^{\overline{ABCDEX}}$ denotes the conditional probability that a person belongs to category (i,j,k,l,m) of the joint manifest variable $ABCDE$, given $X = t$. The parameter $\pi_{it}^{\bar{A}X}$ is the conditional response probability that an individual obtains score $A = i$, given this person belongs to latent class t of X. The remaining conditional response probabilities in Equation 15 have similar meanings. All quantities in Equations 14 and 15 are probabilities and subject to the standard restrictions: They cannot be smaller than 0 or larger than 1 and their sum is 1 after summation over the appropriate subscripts, for example, $\Sigma_t \pi_t^X = \Sigma_i \pi_{it}^{\bar{A}X} = 1$.

Equation 14 implies that the population may be divided into T exhaustive and mutually exclusive latent classes; each person belongs to one and only one latent class. In this sense, Equation 14 expresses the existence of latent variable X.

The assumption of local independence is essential to Equation 15. From the rules of probability theory, it follows that the probability π_{ijklmt}^{ABCDEX} of obtaining score (i,j,k,l,m,t) on the joint variable $ABCDEX$ may be written as the probability π_t^X of belonging to $X = t$ multiplied by the conditional response probability $\pi_{ijklmt}^{\overline{ABCDEX}}$ of obtaining score (i, j,k,l,m) on the joint variable $ABCDE$, given $X = t$. According to the local independence assumption, the scores on the manifest variables are independent of each other within each latent class and so the conditional response probabilities $\pi_{it}^{\bar{A}X}$, $\pi_{jt}^{\bar{B}X}$, $\pi_{kt}^{\bar{C}X}$, $\pi_{lt}^{\bar{D}X}$, and $\pi_{mt}^{\bar{E}X}$ are independent of each other. Therefore, $\pi_{ijklmt}^{\overline{ABCDEX}}$, the joint conditional probability of scoring (i,j,k,l,m) on $ABCDE$, given $X = t$, can be obtained by simply multiplying the separate conditional probabilities $\pi_{it}^{\bar{A}X}$, $\pi_{jt}^{\bar{B}X}$, and so on.

The loglinear representation of the latent class model in Figure 3.1 looks like this:

$$F_{ijklmt}^{ABCDEX} = \eta \tau_i^A \, \tau_j^B \, \tau_k^C \, \tau_l^D \, \tau_m^E \, \tau_t^X \, \tau_{it}^{AX} \, \tau_{jt}^{BX} \, \tau_{kt}^{CX} \, \tau_{lt}^{DX} \, \tau_{mt}^{EX}$$

$$G^{ABCDEX}_{ijklmt} = \theta + \lambda^{A}_{i} + \lambda^{B}_{j} + \lambda^{C}_{k} + \lambda^{D}_{l} + \lambda^{E}_{m} + \lambda^{X}_{t} + \lambda^{AX}_{it} + \lambda^{BX}_{jt} + \lambda^{CX}_{kt} + \lambda^{DX}_{lt} + \lambda^{EX}_{mt}, \tag{16}$$

where the parameters are subject to the usual identifying restrictions: The product of the τ-parameters, multiplied over any subscript, equals 1, and the λ-parameters analogously sum to 0.

In hierarchical loglinear model $\{AX, BX, CX, DX, EX\}$ (Equation 16), the latent variable X is directly related to each of the manifest variables, but, in keeping with the local independence assumption, there are no direct relations between the manifest variables. The expected frequencies F^{ABCDEX}_{ijklmt} for this model are identical to π^{ABCDEX}_{ijklmt} in Equation 15 multiplied by sample size N. The τ-parameters in Equation 16 can be rewritten in terms of the π-parameters at the right-hand side of Equation 15.

Because X is a latent variable, the estimated expected frequencies $\hat{F}$ (and $\hat{\pi}$) for model $\{AX, BX, CX, DX, EX\}$ cannot be found by straightforward application of the iterative procedures mentioned in the previous chapter. The EM-algorithm and particular variants of the Newton/Raphson procedure can be used (Appendix B) (Goodman, 1974a, 1974b; Haberman, 1979, chap. 10, 1988; Hagenaars, 1990, sec. 3.2.2). Once the estimated expected frequencies $\hat{F}^{ABCDEX}_{ijklmt}$ have been obtained, the estimated parameters in Equation 16 can be computed using the same formulas that were used for ordinary loglinear models. By summing $\hat{F}^{ABCDEX}_{ijklmt}$ over the latent variable X, the estimated expected frequencies $\hat{F}^{ABCDEX}_{ijklm+}$ can be compared with the observed frequencies f^{ABCDE}_{ijklm} in the usual manner, by means of the chi-square test statistics L^2 and χ^2, to test the empirical validity of the latent class model.

But there are a few problems. The iterative procedures may not converge to the maximum likelihood estimates $\hat{F}$, but may stop at some local maximum. Rerunning the iterative procedures with different initial estimates is therefore recommended. If the rerun yields a different value of L^2, the lowest one is the maximum likelihood solution (unless still lower values may be found). If the same value of L^2 is found, but the estimated expected frequencies $\hat{F}^{ABCDEX}_{ijklmt}$ and the parameter estimates are different, there is an identifiability problem: No unique solution for the parameter set exists.

A necessary condition for identifiability is that the number of independent unknowns (the parameters to be estimated) does not exceed the number of independent knowns (the cell frequencies in the observed

TABLE 3.2
Basic Latent Class Model for Table 3.1

		X. Political Orientation Class 1	Class 2	$\hat{\lambda}$[a]
A. System Responsiveness	1. Low	.315[b]	.678	AX: −.380
	2. High	.685	.322	
B. Ideological Level	1. Ideologues	.418	.021	BX: .879
	2. Nonideologues	.582	.979	
C. Repression Potential	1. Low	.418	.265	CX: .172
	2. High	.582	.735	
D. Protest Approval	1. Low	.293	.610	DX: −.332
	2. High	.707	.390	
E. Conventional Participation	1. Low	.172	.695	EX: −.599
	2. High	.828	.305	
	Latent class size $\hat{\pi}_t^X$	.600	.400	

$L^2 = 95.80$, df = 20, $p = .00$
$\chi^2 = 95.31$

a. Entries in this column are the values of the loglinear two-variable parameter $\hat{\lambda}_{11}^{AX}$, $\hat{\lambda}_{11}^{BX}$, and so on.
b. Entries in this column and the next are, except for the last row, the conditional response probabilities $\hat{\pi}_{it}^{\bar{A}X}$, $\hat{\pi}_{jt}^{\bar{B}X}$, and so on.

table). However, this is not a sufficient condition. Goodman (1974b) states the sufficient conditions for local identifiability by means of which it can be determined whether there are other solutions for the parameter estimates in the neighborhood of the solution found that give the same L^2 value (see also De Leeuw, Van der Heijden, & Verboon, 1990; Van der Heijden, Mooyaart, & De Leeuw, 1992). Several programs (e.g., MLLSA, NEWTON; see Appendix A) will inform the user about the local identifiability of the model parameters.

For identifiable models, the number of degrees of freedom for testing the fit of a model equals the number of cell frequencies of the observed table minus the number of parameters to be estimated independently.

Table 3.2 shows the relevant parameter estimates for the basic latent class model $\{AX, BX, CX, DX, EX\}$ applied to Table 3.1, assuming that X has only two categories. From the test results mentioned at the bottom of Table 3.2 it must be concluded that the proposed latent class model does not fit the data. However, for illustrative purposes I will discuss the outcomes.

It may be inferred from the estimates $\hat{\pi}_t^X$ that 60% of the people belong to latent Class 1 and 40% to latent Class 2. The conditional response probabilities show that those who belong to latent Class 1 are characterized by a high opinion of System Responsiveness, a low Repression Potential, a high Protest Approval, a high level of Conventional Participation, and a large number of Ideologues, relative to the members of latent Class 2.

The loglinear parameters presented in the last column of Table 3.2 confirm this pattern of association between the latent and the manifest variables. The loglinear parameters $\hat{\lambda}$ rather than their multiplicative counterparts $\hat{\tau}$ have been given here to facilitate the comparison of negative and positive associations. They show that Conventional Participation (*E*) and particularly Ideological Level (*B*) are most strongly related to the latent variable *X* and that Repression Potential (*C*) has the weakest ties with *X*. The odds that someone belongs to the Ideologues ($B = 1$) rather than to the Nonideologues ($B = 2$) are 33.650 times higher for the people from the first ($X = 1$) than for those from the second latent class ($X = 2$). The analogous odds ratio for the relation between the latent variable *X* and Repression Potential (*C*) is only 1.990. These values of the odds ratios can be obtained by first taking the antilogarithm of the $\hat{\lambda}$'s to get the $\hat{\tau}$-parameters and then computing $(\hat{\tau})^4$, because in 2×2 tables, when effect coding is used, $(\hat{\tau}_{1\ 1})^4$ equals the odds ratio.

There is no need to distinguish carefully between partial and marginal odds ratios when discussing the odds ratios for the relations between the latent variable and the manifest variables. According to the collapsibility theorem, because of the special nature of the basic latent class model in which there are no direct relations among the manifest variables, the marginal odds for the relations between the latent variable and a particular manifest variable are the same as the corresponding partial odds, holding the other manifest variables constant. The odds ratios and the loglinear two-variable parameters can thus also be obtained by means of the conditional response probabilities presented in Table 3.2.

As in factor analysis, the meaning of the latent variable *X* follows mainly from the relationships between the latent and the manifest variables. In this case, it is not easy to find a suitable label for the members of latent Classes 1 and 2; the presumed labels "liberal" and "conservative" are certainly not adequate. But, as the model does not fit anyway, we are not bothered by this.

Several variants of the basic latent class model may be obtained by imposing a priori restrictions on the parameters. Indicators can be defined as "perfect" indicators of the latent variable by setting the appropriate conditional response probabilities equal to one or zero. For example, when considering B to be a perfect indicator of X, the restrictions $\pi_{\bar{1}\,\bar{1}}^{\bar{B}X} = \pi_{\bar{2}\,\bar{2}}^{\bar{B}X} = 1$ could be imposed.

It is also possible to define certain manifest variables, such as A and D, as equal, parallel indicators of the latent variable by a priori equating their conditional response probabilities, $\pi_{\bar{1}\,1}^{\bar{A}X} = \pi_{\bar{1}\,1}^{\bar{D}X}$ and $\pi_{\bar{2}\,2}^{\bar{A}X} = \pi_{\bar{2}\,2}^{\bar{D}X}$, or their loglinear "links" with the latent variable, $\lambda_{1\,1}^{AX} = \lambda_{1\,1}^{DX}$. Note that imposing equality restrictions on the conditional response probabilities is generally not the same as imposing equality restrictions on the loglinear two-variable effects. The fact that the parallelism of A and D involved two restrictions in terms of conditional response probabilities but just one restriction in terms of loglinear effects already pointed to that. Conditional response probabilities are a function of loglinear two- and one-variable parameters, and equality restrictions on the conditional response probabilities usually involve restrictions on both two- and one-variable effects (see Chapter 6, Equations 20-27).

For these and other variants of the basic model, the reader should consult not only the references cited at the beginning of this chapter, but also Mooyaart and Van der Heijden (1992) for a correction of some of the generally accepted previous statements about obtaining maximum likelihood estimates in the presence of equality restrictions on the conditional response probabilities.

Modifying the Basic Model: Models With Two or More Latent Variables

As the test statistics in Table 3.2 clearly show, the simple basic latent class model does not fit the data in Table 3.1. There are several tools available to help detect the causes of the bad fit and to decide how the basic model may be improved.

From inspection of the two-variable effects in Table 3.2, it is clear that variable C is rather weakly related to the latent variable X and, therefore, it may be better to remove it. The same two-latent-class model is applied again, but now with the four manifest variables A, B, D, and E. Although a well-fitting model results ($L^2 = 7.88$, df $= 6$, $p = .25$, $\chi^2 = 8.06$), this line of analysis will not be pursued any further because it is still difficult to find a good substantive interpretation of the outcomes.

Another possibility for improving on the original model is to keep all five manifest variables in it, still postulating the existence of one latent variable X, but extending the number of classes of this latent variable. The test results for the three-latent-class model are on the borderline: $L^2 = 24.943$, df = 14, $p = .035$ ($\chi^2 = 24.638$). It is still hard to find a good substantive interpretation of the outcomes. Models with four and more latent classes have not been tried. In general, one must be careful with extending the number of latent classes without imposing additional restrictions. The resulting models may not be identified, even if there seem to be sufficient degrees of freedom. For example, the basic latent class model with four dichotomous manifest variables and one trichotomous latent variable is not identified, even though the number of independent parameters seems to be one less than the number of knowns. Additional restrictions are needed to make it identifiable.

The major weak point of the two-latent-class model presented in Table 3.2 is that there are only very vague ideas about the nature of the possible underlying variable X. A closer and more critical look at the contents of the five manifest variables suggests that perhaps the label "political orientation," in the sense of liberal versus conservative, might be appropriate for the Protest Tolerance items C (Repression Potential) and D (Protest Approval), but not for the other three. Variables A, B, and E indicate how involved the respondents are in the present political system, regardless of their left-right inclinations: Item A registers the respondent's positive or negative evaluation of the system's responsiveness to the needs of the citizens; Item B, the respondent's understanding of the system in political rather than idiosyncratic terms; Item E, the degree of active participation in the system. So perhaps variables A, B, and E are indicators of a different concept from that indicated by C and D, and maybe we need two latent variables to explain the associations among the five indicators.

Before testing the validity of these substantive considerations by setting up a two-latent-variable model, it is worth noting that there are several indications in the data that point in the same direction. First, there is the evidence from the pattern of observed associations among the indicators. The observed marginal tables AB, AC, . . . , DE can be computed by means of Table 3.1, and for each marginal table the loglinear two-variable parameters of the saturated models $\{AB\}$, $\{AC\}$, . . . , $\{DE\}$ can be calculated. Inspection of these coefficients suggests two clusters of variables, A-B-E and C-D, which have rather high within but very low between associations.

Then there are the residual frequencies of the two-latent-class model of Table 3.2. The standardized residuals

$$r_{i\,j\,k\,l\,m}^{ABCDE} = (f_{i\,j\,k\,l\,m}^{ABCDE} - \hat{F}_{i\,j\,k\,l\,m+}^{ABCDEX})/\sqrt{\hat{F}_{i\,j\,k\,l\,m+}^{ABCDEX}}$$

were inspected. Because it is very hard to discover with the naked eye a systematic pattern in the sizes and signs of these 32 residuals, marginal residual tables $r_{i\,j+++}^{ABCDE}$, r_{i+k++}^{ABCDE}, . . . , r_{+++lm}^{ABCDE} were set up (Hagenaars, 1988a). It then became apparent that the entries for residual table *CD* were by far the largest and that the observed frequencies $f_{1\,1}^{CD}$ and $f_{2\,2}^{CD}$ were seriously overestimated by $\hat{F}_{1\,1}^{CD}$ and $\hat{F}_{2\,2}^{CD}$, respectively, whereas the observed frequencies f for cells (1,2) and (2,1) were clearly underestimated by $\hat{F}$.

Obviously, the observed marginal association *CD* is not well reproduced by the basic latent class model. If the observed marginal table *CD* with frequencies $f_{k\,l}^{CD}$ is used, $\hat{\lambda}_{1\,1}^{CD}$ turns out to be −.308. Setting up the estimated expected marginal table *CD* by means of the estimated expected frequencies $\hat{F}$ obtained for the two-latent-class model in Table 3.2, the value of $\hat{\lambda}_{1\,1}^{CD}$ is a mere −.052. Carrying out the same exercise for the other marginal two-way associations among the manifest variables reveals no further important differences between the observed and the estimated expected marginal associations. Only the strength of the observed marginal association *CD* is seriously underestimated by the two-latent-class model. Apparently, there are additional sources of association between *C* and *D*, other than *X*, possibly another latent variable affecting *C* and *D* (for other conceivable explanations, see Hagenaars, 1988a).

The basic latent class model may be modified into a two-latent-variable model {*YZ, AY, BY, EY, CZ, DZ*}, where the latent variable *Y* refers to the presumed underlying variable System Involvement with indicators *A*, *B*, and *E*, and the latent variable *Z* refers to the latent variable Protest Tolerance with indicators *C* and *D*. The path diagram representing this two-variable latent class model is depicted in Figure 3.2.

The parameters of model {*YZ, AY, BY, EY, CZ, DZ*} were estimated for the data in Table 3.1 assuming that both *Y* and *Z* are dichotomous. The outcomes are presented in Table 3.3. As the test results show, this two-latent-variable model fits the data very well. And, equally important, the outcomes lend themselves to a meaningful interpretation.

Looking at this latent class model from the loglinear perspective, the model is a hierarchical loglinear model with two dichotomous latent variables *Y* and *Z* in which *A*, *B*, and *E* are directly influenced only by *Y*, and *C* and *D* directly only by *Z*. Estimates of the sizes of these direct

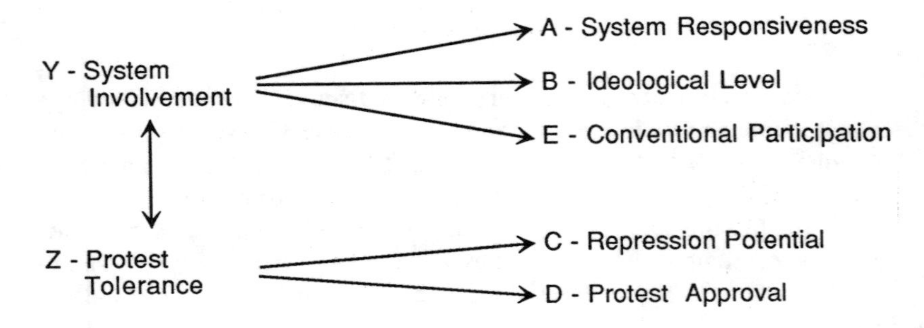

Figure 3.2. Two-Latent-Variable Model for Table 3.1
NOTE: Manifest variables, *A, B, C, D, E*; latent variable, *Y, Z*.

influences are presented in the last column of Table 3.3. *D* and *Z* in particular are very strongly related: *D* is an almost perfect indicator of Z.

From the signs of these effects it can be inferred that people belonging to the first latent class of *Y* have a higher involvement in the present political system than the respondents in $Y = 2$. The people in latent class $Y = 1$ have a higher opinion of the system's responsiveness, there are more ideologues among them, and they engage in more political activities, compared with the people in latent class 2 of *Y*. Those who belong to $Z = 1$ are more tolerant toward protests than the respondents from latent class 2 of *Z*. The people from latent class $Z = 1$ have a lower repression potential and are definitely more lenient with regard to forms of protest than the people in $Z = 2$.

The last figure in the last column of Table 3.3 shows the association between the two latent variables: $\hat{\lambda}_{1\,1}^{YZ} = .286$, indicating a partial (and marginal—see below) odds ratio of 3.139 [$= (e^{.286})^4 = 1.3311^4$]. This is a moderate positive relationship. There is no immediately clear asymmetrical causal order between *Y* and *Z*. The odds ratio implies that the odds "high/low Protest Tolerance" are 3.139 times larger among those highly involved in the political system than among those less involved. But it is just as meaningful to conclude that the odds "high/ low System Involvement" are 3.139 times larger for the people who score high on Protest Tolerance than for the more intolerant people. Both interpretations of the odds ratio make sense and both are formally correct. Odds ratios as well as higher-order odds ratios and loglinear parameters are in this sense symmetric coefficients (Hagenaars, 1990, pp. 30-31).

Looking at the two-latent-variable model from the standpoint of the parametrization in terms of probabilities (Equation 15), we have a

TABLE 3.3
Two-Latent-Variable Model for Table 3.1

Y. System Involvement		*Class 1* 1. High	*Class 2* 1. High	*Class 3* 2. Low	*Class 4* 2. Low	
Z. Protest Tolerance		1. High	2. Low	1. High	2. Low	$\hat{\lambda}$[a]
A. System	1. Low	.306[b]	.306	.684	.684	AY: −.399
Responsiveness	2. High	.694	.694	.316	.316	
B. Ideological	1. Ideologues	.415	.415	.033	.033	BY: .757
Level	2. Nonideologues	.585	.585	.967	.967	
E. Conventional	1. Low	.147	.147	.719	.719	EY: −.674
Participation	2. High	.853	.853	.281	.281	
C. Repression	1. Low	.483	.204	.483	.204	CZ: .324
Potential	2. High	.517	.796	.517	.796	
D. Protest	1. Low	.000	.929	.000	.929	DZ: −2.786[c]
Approval	2. High	1.000	.071	1.000	.071	
Latent class size $\hat{\pi}_{rs}^{YZ}$		.392	.200	.157	.251	YZ: .286

$L^2 = 19.042$, df = 18, $p = .39$
$\chi^2 = 19.264$

a. Entries in this column are the values of the loglinear two-variable parameter $\hat{\lambda}_{11}^{AY}$, $\hat{\lambda}_{11}^{CZ}$, and so on.
b. Entries in this column and the next three are, except for the last row, the conditional response probabilities $\hat{\pi}_{irs}^{\bar{A}YZ}$, $\hat{\pi}_{jrs}^{\bar{B}YZ}$, and so on; identical values are the result of a priori equality restrictions.
c. If the conditional response probabilities .000 and 1.000 had been exactly 0 and 1, this loglinear parameter would not have been defined. See text for further comments on this.

four-latent-class model in which the four latent classes result from the cross-classification of the two dichotomous latent variables Y and Z (Goodman, 1974a, 1974b; Lazarsfeld & Henry, 1968, p. 90). As the structure of this four-latent-class model is such that A, B, and E depend only on latent variable Y, and C and D only on latent variable Z, equality restrictions have been imposed on the conditional response probabilities to ensure that the response probabilities concerning A, B, and E vary only with Y, and C and D only with Z. The thus restricted latent class model is exactly equivalent with model $\{YZ, AY, BY, EY, CZ, DZ\}$, and the estimated probabilities $\hat{\pi}_{ijklmrs}^{ABCDEYZ}$ multiplied by N are identical to the estimated expected frequencies of model $\{YZ, AY, BY, EY, CZ, DZ\}$.

The entries $\hat{\pi}_{rs}^{YZ}$ of themarginal 2×2 table YZ are presented in the last row of Table 3.3. They can be used to obtain $\hat{\lambda}_{11}^{YZ}$, with the same numerical result as before (.286). Given the nature of loglinear model $\{YZ, AY, BY, EY, CZ, DZ\}$, in which none of the manifest variables is directly related to both latent variables, it follows from application of the collapsibility theorem that the partial and marginal odds indicating the association between Y and Z have the same value.

It is interesting to compare this value with the corresponding values of the observed marginal two-variable associations between the manifest variables A, B, E, on the one hand, and C, D, on the other. The average association based on the absolute values $|\hat{\lambda}_{1\,1}|$ for AC, AD, BC, BD, CE, and DE equals .084, the strongest one being −.178 for BD. The relation between the latent variables is stronger than the corresponding relations between the manifest variables. So we observe here a phenomenon akin to the correction for attenuation caused by unreliability of the measurements, well known from classical test theory (Bohrnstedt, 1983).

Finally, the estimated values of the conditional response probabilities for D deserve special attention. When $Z = 1$, it is absolutely certain that people will score high on Protest Approval and, consequently, there is no chance that they will respond low on this item. When using the EM-algorithm (Appendix B) to obtain the maximum likelihood estimates of the parameter estimates, as was done here, it is impossible to get estimated probabilities smaller than zero or larger than one. If during the iterative procedure these boundaries are reached for a particular estimated probability, the value for this probability will not change any more during successive iterations. The obtained solution for the whole parameter set should then be regarded as a terminal solution, or as a conditional maximum likelihood solution, conditional upon the fact that these boundary estimates reflect the true values in the population, that is, as if these restrictions were imposed a priori. It remains open to debate whether or not the degrees of freedom have to be adjusted for this posterior employment of an a priori restriction on the parameters concerned.

To the extent that the conditional response probabilities .000 and 1.000 reported for D approach the exact values 0 and 1, respectively, $\hat{\lambda}_{1\,1}^{DZ}$ approaches the limiting value $-\infty$. Because .000 and 1.000 were not exactly 0 and 1, respectively, and because the calculations were carried out with greater precision than the three decimals reported in Table 3.3, the value of −2.786 could be obtained for $\hat{\lambda}_{1\,1}^{DZ}$, the absolute value of which implies an odds ratio of as much as 69147.695.

External Variables and Latent Class Scores

Once a satisfactory latent class model has been found, the investigator often wants to continue the analysis and relate the newfound latent variables to certain external variables by cross-tabulating the external variables with the latent variable. For example, it might be interesting to find out what kinds of people in terms of Age, Sex, or Education have high or low System Involvement. Obviously, there is a difficulty here because System Involvement is not a directly observed variable.

Factor analysis presents the same problem. A common solution is to compute an "observed" factor score for each individual and continue the analysis by correlating these factor scores with the scores on the relevant external variables. Analogously, "observed" latent class scores may be obtained for the latent class model. The starting point for the computation of the "observed" latent class scores is the classification probability, which is the conditional probability that a person belongs to latent class $X = t$, given the scores on the manifest variables:

$$\hat{\pi}_{i\,j\,k\,l\,m\,t}^{ABCDE\bar{X}} = \hat{\pi}_{i\,j\,k\,l\,m\,t}^{ABCDEX} / \hat{\pi}_{i\,j\,k\,l\,m\,+}^{ABCDEX} . \tag{17}$$

Equation 17 and the maximum likelihood estimates $\hat{\pi}$ can be used to estimate the conditional probability that a person with observed scoring pattern (i,j,k,l,m) on variables A through E belongs to latent class $t = 1, \ldots, T$. The classification probability can be used in several ways to assign respondents to one of the latent classes. Usually, the modal assignment rule is employed: A person will be assigned to that latent class for which the estimated classification probability is largest, given the manifest scoring pattern.

Although "observed" latent class scores are certainly useful, two major problems are related to their use. (Many researchers seem not to be aware of it, but two analogous problems are connected with the use of common factor scores; see Saris, De Pijper, & Mulder, 1978; Steiger, 1979a, 1979b.) The first problem is that the "observed" latent class scores are not perfect substitutes for the true individual latent class scores. Except for the extreme cases where the estimated classification probabilities for a given scoring pattern are equal to one or zero, there is a chance equal to one minus the modal classification probability that the individual does not belong to the modal class. Therefore, most probably, several individuals will be assigned to the wrong latent class, with the result, among other things, that the true relations between latent and external variables are not correctly reflected by the relations between the "observed" latent class scores and external variables (Clogg, 1981b, sec. 5; Hagenaars, 1990, pp. 113-119, 142; McCutcheon, 1987, pp. 35-37).

The other problem with "observed" latent class scores, perhaps more difficult to recognize, is that the true individual latent class scores themselves are not identified within the latent class model. Even when one knows all observed scores for each individual as well as the true values of the latent class parameters $\pi_{i\,j\,k\,l\,m\,t}^{ABCDEX}$ in the population, it is still not possible to determine exactly what the true score on the latent variable X is for each individual. Several different sets of "true" individual X scores may be assigned to the individuals, all in agreement with the values of the model parameters and with the observed scores, but not identical to each other and in some cases negatively correlated with each other (Hagenaars, 1990, pp. 115-116). It is doubtful what the usefulness is of getting "observed" estimates of unidentified individual true scores.

How severe these two problems are depends mainly on the strength of the relations between the latent and the manifest variables. In the extreme case that one or more manifest variables are perfectly related to the latent variable and can be used as identical substitute(s) for the latent variable, there are no misclassifications and the problem of unidentifiability of the individual latent class scores disappears. But otherwise, these problems are severe enough that one should look for an alternative way of determining the relations between latent and external variables.

The alternative is to make the external variables an internal part of a structural model with latent variables and to obtain simultaneously estimates for the relationships among all three kinds of variables: latent, manifest, and external. For interval-level variables, this has become standard routine with the introduction of programs such as LISREL and EQS. For categorical measurement, an analogous solution is possible. In the next two chapters, a "modified LISREL approach" will be presented in which the external variables are incorporated into loglinear models with latent variables (Hagenaars, 1990, sec. 3.4.2).

In conclusion, it should be kept in mind that although all examples presented here concerned dichotomous variables in order to keep the explanation of the basic principles of latent class analysis as simple as possible, latent class analysis can handle polytomous variables equally well. Furthermore, no linearity of the relationships among the manifest and latent variables is presumed in latent class analysis, as opposed to factor analysis. There also need not be a one-to-one correspondence between the number of categories of the latent and the manifest variables. Clogg (1981a), Van der Heijden et al. (1992), and the references cited above present many examples illustrating the usefulness of these features.

4. LOGLINEAR MODELING WITH LATENT VARIABLES: INTERNALIZING EXTERNAL VARIABLES

Once the latent class model is recognized as a loglinear model with latent variables, introducing additional external variables into the loglinear model is a logical next step. Computer programs for handling such loglinear models with manifest, latent, and external variables are mentioned in Appendix A.

As an illustration, the data in Table 4.1 are used to answer questions concerning whether men and women are differently involved in the political system and whether this possible difference varies according to age. Many reasons could be cited why men and women might have different relations with the political system. Owing to differences in education and upbringing, men and women may have learned to play different social roles, in which women act mostly in the private and men in the public domain. As social roles are defined differently for different stages of the life course and as role definitions may change from generation to generation, sex differences may be expected to vary with age. More specifically, it is hypothesized that women are less involved in the political system than men, but that this difference is smaller in the younger age groups.

TABLE 4.1

Indices of System Involvement (A, B, E) by Sex (S) and Age (G)

		S. Sex	*1. Men*			*2. Women*			
		G. Age	*1.*	*2.*	*3.*	*1.*	*2.*	*3.*	*Total*
A. System Respon-siveness	*B.Ideological Level*	*E. Conventional Participation*	*16-34*	*35-57*	*58-91*	*16-34*	*35-57*	*58-91*	
1. Low	1. Ideologues	1. Low	5	4	6	5	3	1	24
		2. High	15	22	10	9	18	6	80
	2. Nonideologues	1. Low	31	25	27	47	49	65	244
		2. High	24	36	21	29	40	34	184
2. High	1. Ideologues	1. Low	7	5	2	7	1	6	28
		2. High	26	38	29	23	29	23	168
	2. Nonideologues	1. Low	24	11	14	38	21	36	144
		2. High	51	47	28	34	59	65	284
		Total	183	188	137	192	220	236	1165

SOURCE: See Table 2.1.

a.

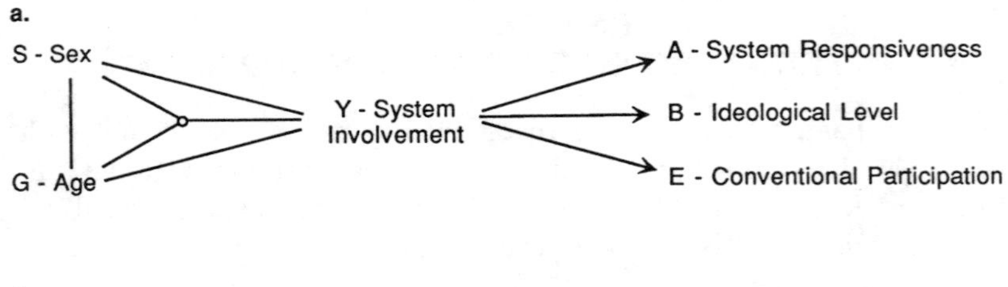

b.

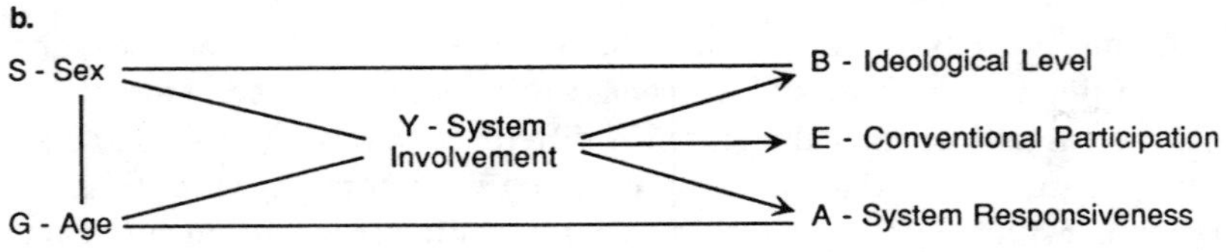

Figure 4.1. Loglinear Models With External, Manifest, and Latent Variables for Table 4.1
NOTE: Manifest variables, *A, B, E*; latent variable, *Y*; external variables, *S, G*.

First, to test whether the three indicators—System Responsiveness (*A*), Ideological Level (*B*), and Conventional Participation (*E*)—taken by themselves can still be regarded as indicators of System Involvement (*Y*), a two-latent-class analysis was carried out on the row totals of Table 4.1, which contain the observed frequencies of marginal table *ABE*. With three dichotomous manifest variables and one dichotomous latent variable, model {*AY, BY, EY*} is exactly identified, has zero degrees of freedom, and fits the data perfectly ($L^2 = \chi^2 = 0$). The parameter estimates concerning the relations between latent variable *Y* and indicators *A, B,* and *E* are very similar to those mentioned in Table 3.3, and the latent variable *Y* may be interpreted as System Involvement.

To test the hypotheses about the differences between men and women with regard to their system involvement, models must be defined that connect Sex (*S*) and Age (*G*, for *generation*) with the latent variable *Y*. An obvious model, given the hypotheses, is model {*SGY, AY, BY, EY*} depicted in Figure 4.1a. According to this model, Sex and Age directly influence latent variable *Y* (System Involvement) and also affect each other's influence on *Y*. The relations between the external variables and the indicators are only indirect through *Y*.

The test statistics are presented in Table 4.2, Model 1. The *p* value is on the borderline. It is doubtful whether this model should be accepted,

TABLE 4.2
Test Results for Table 4.1

Model	L^2	df	*p*	χ^2
1. {*SGY,AY,BY,EY*}	49.382	30	.014	46.970
2. {*SG,SY,GY,AY,BY,EY*}	49.595	32	.024	47.364
3. {*SG,GY,AY,BY,EY*}	72.174	33	.000	69.106
4. {*SG,SY,AY,BY,EY*}	63.313	34	.002	60.833
5. {*SG,AY,BY,EY*}	88.408	35	.000	84.241
6. {*SG,SY,GY,SB,GA,AY,BY,EY*}	29.287	29	.450	27.102
7. (2) - (1)[a]	.213	2	.899	
8. (3) - (2)	22.579	1	.000	
9. (4) - (2)	13.718	2	.000	
10. (5) - (2)	38.813	3	.000	
11. (2) - (6)	20.308	3	.000	

NOTE: *S* = Sex; *G* = Age; *A* = System Responsiveness; *B* = Ideological Level; *E* = Conventional Participation; *Y* = System Involvement (latent variable).
a. Models 7 through 11 provide the conditional test outcomes of $L^2_{r/u}$ for the restricted and unrestricted models mentioned.

but for the time being it will be. Looking at the parameter estimates with regard to the relations between *Y* and its indicators *A, B,* and *E* (not reported here), they appear to be very similar to those obtained for the two-latent-class model {*AY, BY, EY*} applied to the row totals of Table 4.1.

The estimates for those parameters that are most relevant for the hypotheses we started with are $\hat{\lambda}^{SY}_{11} = .203$, $\hat{\lambda}^{GY}_{11} = -.127$, $\hat{\lambda}^{GY}_{21} = .232$, $\hat{\lambda}^{GY}_{31} = -.105$, and $\hat{\lambda}^{SGY}_{111} = .017$, $\hat{\lambda}^{SGY}_{121} = .011$, $\hat{\lambda}^{SGY}_{131} = -.028$ (where the categories of *Y* are 1, high, and 2, low). In agreement with our hypothesis, men appear to have a higher degree of System Involvement than women; the partial odds "high/low System Involvement" are 2.252 [= $(e^{.203})^4$] times higher for men than for women. People from the middle age category, 35-57, are more involved than both the younger and the older; the partial odds "high/low System Involvement" are about two times higher for the middle age group than for the other two age groups.

The figure of two times higher is obtained as follows. First, the antilogarithms of the parameter estimates $\hat{\lambda}^{GY}_{11}$, and so on, are taken. It then becomes clear that on average the estimated expected frequencies $GY = (2,1)$, $\hat{F}^{ABESGY}_{ijkl21}$, are 1.261 (= $e^{.232}$) higher than expected on the basis of the lower-order effects, whereas the estimated expected frequencies for which $GY = (2,2)$ are .793 (= 1/1.261) times lower. So, for the middle age category, according to the two variable effects, the partial odds "high/low System Involvement" are 1.590 (= 1.261/.793) times higher than expected on the basis of the lower-order effects. The corresponding figures for the youngest and oldest age groups are .776 and .811, respectively. So, for the middle age category, the partial odds "high/low System Involvement" are 2.049 (= 1.590/.776 ≈ 2) times higher than for the youngest age group and 1.961 (= 1.590/.811 ≈ 2) times higher than for the oldest group.

The three-variable interaction parameter estimates are very small. Contrary to our hypothesis, the man/woman differences with regard to System Involvement are the same in all age categories (and the age differences are the same for men and women). Not surprisingly, deleting these three-variable effects from the model does not significantly worsen the fit (Models 2 and 7 in Table 4.2), nor does it change in any meaningful way the parameters estimates for the relations *S*-*Y* and *G*-*Y*. When the direct effects *S*-*Y* or *G*-*Y* are also deleted, the fit becomes significantly worse (Models 3, 4, 5, and 8, 9, 10 in Table 4.2). Sex and Age appear to have a statistically significant influence on System Involvement.

By means of these latent variable models in which external variables are incorporated, an answer to the substantive question has been obtained, albeit not a completely satisfactory one. The test outcomes for the final Model 2 (Table 4.2) are on the borderline. Inspection of the standardized residuals of Model 2 in the manner described above suggests that the model might be improved by adding the direct effects of Sex on *B* (Ideological Level) and of Age on *A* (System Responsiveness). The resulting model, depicted in Figure 4.1b, fits the data very well (Model 6, Table 4.2) and much better than Model 2 (Model 11, Table 4.2).

The extra parameter estimates *BS* and *AG* (not reported here) are not very large, but evidently statistically significant. Their interpretation poses some problems. It is not immediately clear how to interpret the direct effects of external variables on manifest variables where the latter variables are meant only to be indicators of latent variables. Such direct effects may imply that the indicators have a different meaning for different subgroups. In this case, it may be that the meaning of *B* (Ideological Level) is a bit different for men and women and the meaning of *A* (System Responsiveness) differs somewhat according to Age. In terms of test theory, direct effects between external variables and indicators point to the possibility of "test item bias" (Osterlind, 1983).

I will not go into this any further here. The main purpose of introducing Model 6 was to show that careful inspection of the residuals may reveal what is wrong with a particular model, including possible shortcomings of the measurement model.

When employing models such as those in Figure 4.1, one has to be sure that no asymmetrical, causal effects are intended. Otherwise, more often than not such models have to be tested and their parameters estimated in the stepwise manner of modified path models described in Chapter 2. I have not dealt with this aspect here because it will be introduced explicitly in Chapter 5.

5. CAUSAL MODELS WITH LATENT VARIABLES: A MODIFIED LISREL APPROACH

Loglinear models in the form of modified path models can be used to investigate causal relationships among observed categorical variables (see Chapter 2). Extension of the modified path analysis approach toward a modified LISREL approach by introducing latent variables into the path models follows naturally (Hagenaars, 1990, sec. 3.7). The modified LISREL approach is illustrated in this chapter, using the data in Table 5.1.

Table 5.1 is a large and complex table with many variables. Many of its cell entries are zero or very small. A table such as this generally presents investigators with a number of problems. Nevertheless, it presents the kind of data investigators often work with. The loglinear analysis of this table is complicated, as will be the exposition of the analyses, making this chapter more difficult than the other chapters in this book. At the end of the chapter, however, a very clear picture of the relations among the variables emerges. The final model chosen for Table 5.1 is depicted in Figure 5.1. It may be helpful to keep this figure in mind.

At the start of the analysis, the causal order among the variables has to be established. In the modified LISREL approach, the causal relations among the variables have to be unidirectional (although Mare & Winship, 1991, have made the first ingenious attempts toward the development of nonrecursive causal loglinear models with reciprocal effects). The variables in which we are interested are the external variables Sex (S), Education (T, for *training*), and Age (G, for *generation*); the two dichotomous latent variables, System Involvement (Y) and Protest Tolerance (Z); and indicators A through D. Sex (S) and Age (G) are

TABLE 5.1

Indices of System Involvement (*A, B, E*), Protest Tolerance (*C, D*) and External Variables Sex (*S*), Age (*G*), and Education (*T*)

					S	1	1	1	1	1	1	2	2	2	2	2	2	
					T	1	1	1	2	2	2	1	1	1	2	2	2	
					G	1	2	3	1	2	3	1	2	3	1	2	3	T
A	B	*C*	*D*	*E*														
1	1	1	1	1		0	0	0	0	0	1	0	0	1	0	1	0	3
1	1	1	1	2		0	1	0	0	1	0	0	0	0	0	1	0	3
1	1	1	2	1		4	0	0	1	0	1	1	0	0	2	1	0	10
1	1	1	2	2		6	3	1	3	1	1	4	4	1	1	1	0	26
1	1	2	1	1		0	0	1	0	2	1	0	0	0	0	0	0	4
1	1	2	1	2		2	3	0	0	2	2	0	3	2	0	4	1	19
1	1	2	2	1		0	1	1	0	1	1	2	1	0	0	0	0	7
1	1	2	2	2		4	7	4	0	4	2	4	2	0	0	3	2	32
1	2	1	1	1		0	1	0	8	1	2	1	0	0	5	3	7	28
1	2	1	1	2		0	1	0	3	5	3	0	0	1	1	2	2	18
1	2	1	2	1		2	0	1	9	4	2	8	3	0	12	6	1	48
1	2	1	2	2		10	3	1	6	2	1	7	3	1	11	6	3	54
1	2	2	1	1		0	1	0	2	10	13	3	2	2	8	21	47	109
1	2	2	1	2		0	3	1	1	11	10	2	1	5	4	13	17	68
1	2	2	2	1		1	1	2	9	7	7	3	3	0	7	11	8	59
1	2	2	2	2		1	4	0	3	7	5	1	9	1	3	6	4	44
2	1	1	1	1		0	0	0	1	0	1	0	0	0	1	0	0	3
2	1	1	1	2		1	3	2	0	0	2	0	3	0	0	0	1	12
2	1	1	2	1		2	1	0	0	0	0	2	1	0	1	0	1	8
2	1	1	2	2		11	6	5	2	4	0	13	7	4	1	1	1	55
2	1	2	1	1		1	1	0	0	0	1	1	0	0	0	0	3	7
2	1	2	1	2		1	5	10	1	0	5	0	2	4	1	4	5	38
2	1	2	2	1		3	2	0	0	1	0	2	0	2	0	0	0	10
2	1	2	2	2		10	17	4	0	3	1	5	7	4	3	5	4	63
2	2	1	1	1		0	0	0	3	0	1	0	2	0	4	2	4	16
2	2	1	1	2		1	3	0	0	2	1	1	1	3	2	0	2	16
2	2	1	2	1		2	1	0	7	0	3	6	0	1	8	4	1	33
2	2	1	2	2		13	7	1	10	4	3	9	8	6	4	6	9	80
2	2	2	1	1		0	1	0	3	4	6	4	0	6	6	4	15	49
2	2	2	1	2		5	7	4	2	7	8	2	6	9	3	15	24	92
2	2	2	2	1		6	1	1	3	4	3	3	4	4	7	5	5	46
2	2	2	2	2		14	8	6	6	9	5	7	6	8	6	17	4	96
					Total	83	96	92	100	92	45	101	142	171	91	78	65	1156

SOURCE: See Table 2.1.

NOTE: *A* = System Responsiveness (1, low; 2, high); *B* = Ideological Level (1, ideologues; 2, nonideologues); *C* = Repression Potential (1, low; 2, high); *D* = Protest Approval (1, low; 2, high); *E* = Conventional Participation (1, low; 2, high); *S* = Sex (1, men; 2, women); *T* = Education (training) (1, some college; 2, less than college); *G* = Age (generation) (1, 16-34; 2, 35-57; 3, 58-91).

considered to be the ultimate independent variables, the exogenous variables of the modified LISREL model with an unspecified correlation between them. *S* and *G* may influence Education (*T*). *S, G,* and *T* may affect *Y*, the latent variable System Involvement with indicators *A, B,* and *E*. Variables *S, G, T,* and *Y* may influence *Z*, the latent variable Protest Tolerance with indicators *C* and *D*. As noted in Chapter 3, the direction of the causal relation between *Y* and *Z* is somewhat arbitrary. For our purposes, System Involvement is considered to be a more fundamental and general characteristic than Protest Tolerance.

In this example, all external variables happen to act as causes of the latent variables. In general, external variables may be causes as well as consequences of latent variables (see, for example, Hagenaars, 1990, p. 139).

It follows from the assumed causal order of the variables that the relation between Sex and Age has to be determined by means of marginal table *SG*. Marginal table *SGT* is used to investigate the effects on Education, marginal table *SGTY* is needed for the effects on System Involvement, and marginal table *SGTYZ* for the effects on Protest Tolerance. If all variables had been directly observed variables, models would be fitted to each separate marginal table *SG, SGT, SGTY,* and *SGTYZ* in the manner described in Chapter 2 to obtain the total modified path model. But *Y* and *Z* are latent variables and *SGTY* and *SGTYZ* are not observed tables. The relations among all variables—external, latent, and manifest—have to be estimated simultaneously, taking, however, the assumed causal order into account. The variant of the EM-algorithm to be used is described in Appendix B.

The model-fitting procedure is started with a modified LISREL model in which the saturated model is postulated for each marginal table just mentioned and in which *A, B,* and *E* are directly influenced only by *Y*, and *C* and *D* only by *Z*. The estimated expected frequencies $\hat{F}^*$ of this modified LISREL model are identical to the estimated expected frequencies $\hat{F}$ of model $\{SGTYZ, AY, BY, EY, CZ, DZ\}$. Because the saturated model is postulated for the relations among the external and latent variables, the estimated expected frequencies $\hat{F}^*$ need not be computed in the stepwise manner indicated in Equation 11 (in Chapter 2), in spite of the asymmetrical causal relations among the variables.

That no stepwise estimation procedure is needed may become clear by returning to Equation 11 and Figure 2.1b. The expected frequencies F_{ijk}^{ASE*} for the path model in Figure 2.1b have to be estimated according to the formula in Equation 11, combining the estimated expected frequencies $\hat{F}_{ij}^{AS}$ of model $\{A,S\}$ applied to marginal table *AS* and the

TABLE 5.2
Test Results for Table 5.1

Model	L^*	df*	p	χ^2
1. {*SGTYZ,AY,BY,EY,CZ,DZ*}	396.655	326	.004	395.742
2. {*sgt* {*SG,ST,GT*}, total {*SGTYZ,AY,BY,EY, CZ,DZ*}}	397.904	328	.005	396.713
3. {*sgt* {*SG,ST,GT*}, *sgty* {*SGT,STY,GY*}, total {*SGTYZ,AY,BY,EY, CZ,DZ*}}	399.278	334	.008	398.861
4. {*sgt* {*SG,ST,GT*}, *sgty* {*SGT,STY,GY*}, total {*SGTY,TZ,GZ,YZ,AY, BY,EY,CZ,DZ*}}	424.942	353	.005	418.197
5. {*sgt* {*SG,ST,GT*}, *sgty* {*SGT,STY,GY*}, *sgtyabe* {*SGTY,AY,BY, EY,GE*}, *sgtyzabe* {*SGTYABE, YZ,TZ,GZ*}, total {*SGTYZABE, CZ,DZ,GC*}}	373.626	349	.175	360.573
6. (2) - (1)	1.249	2	.536	
7. (3) - (1)	2.623	8	.956	
8. (3) - (2)	1.374	6	.967	
9. (4) - (1)	28.287	27	.396	
10. (4) - (2)	27.038	25	.354	
11. (4) - (3)	25.664	19	.140	
12. (4) - (5)	51.32	4	.000	

NOTE: If a particular modified path model has to be represented by several submodels, the lowercase letters before the submodel specification denote the marginal table to which the submodel pertains; *total* refers to the complete table *SGTYZABCDE*.

estimated expected frequencies $\hat{F}^{ASE}_{ijk}$ of model {*AS,AE,SE*} applied to table *ASE*. However, when the saturated model {*AS*} is postulated for table *AS*, the estimated expected frequencies $\hat{F}^{AS}_{ij}$ of model {*AS*} are identical to the observed frequencies f^{AS}_{ij} and, as follows from Equation 11, the estimated expected frequencies $\hat{F}^{*ASE}_{ijk}$ are identical to the estimated expected frequencies $\hat{F}^{ASE}_{ijk}$ of model {*AS, AE, SE*}.

The test statistics for model {*SGTYZ, AY, BY, EY, CZ, DZ*} applied to Table 5.1 are reported in Table 5.2, Model 1. Obviously, Model 1 does not fit. However, it does contain a lot of higher-order interactions among *S, G, T, Y,* and *Z* that are certainly not expected to exist in the population, and thus degrees of freedom are wasted. Moreover, many estimated expected cell frequencies are very small. There are indications that especially L^2 is then highly conservative, that is, strongly biased in the direction of falsely rejecting the model (Agresti, 1990, p. 247). So we have to be careful not to give too much weight to the test outcomes (without implying that the test results are actually in favor of Model 1).

The parameter estimates for Model 1 concerning the relations between the latent variables and their indicators (not presented here) follow the pattern found in Table 3.3. The largest discrepancy between the two sets of outcomes concerns the relation between Z and D. $\hat{\lambda}_{1\,1}^{DZ}$ now equals −.826 and not the value −2.786 found before. The latent variables Y and Z can be given the expected interpretations of System Involvement and Protest Tolerance, respectively.

In order to arrive at a more parsimonious model for the relations among the external and latent variables, the estimated expected frequencies $\hat{F}_{i\,j\,k\,r\,s\,l\,m\,n\,o\,p}^{SGTYZABCDE}$ of Model 1 were used to get a first impression of the sizes of the effects among the external and the latent variables. Because Sex and Age are regarded as exogenous variables, their relation in marginal table SG is taken for granted.

The next marginal table to be considered is table SGT. The effects of Sex (S) and Age (G) on Education (T) in marginal table SGT can be estimated using the estimated expected frequencies $\hat{F}_{i\,j\,k+++++++}^{SGTYZABCDE}$ obtained for Model 1. In agreement with previous analyses in Chapter 2, it is found that Sex and Age influence Education, but that the three-variable interaction terms are very small (see Table 2.3). Exclusion of these three-variable interaction terms without imposing any further constraints leads to a modified LISREL model in which $\hat{F}^*$ has to be obtained in a stepwise manner, analogous to Equation 11. The estimated expected frequencies $\hat{F}^*$ for the amended modified LISREL model have to be such that the entries in table SGT, $\hat{F}^{*SGTYZABCDE}_{i\,j\,k+++++++}$, are in conformity with submodel $\{SG,ST,GT\}$, whereas the entries in the complete table satisfy submodel $\{SGTYZ,AY,BY,EY,CZ,DZ\}$. The latter submodel contains the three-variable interaction term SGT because, as explained in Chapter 2, in each submodel the relations among the variables from causally prior submodels are taken for granted. LCAG and other programs mentioned in Appendix A perform the necessary calculations using the EM-algorithm of Appendix B.

The test results for this amended model are reported in Table 5.2, Model 2. Model 2 fits the data no better than Model 1, but definitely no worse. Comparing Models 1 and 2 (Model 6, Table 5.2) yields a nonsignificant conditional test statistic L^* (which is identical to the nonconditional test statistic L^2 obtained above for model {*SG,ST,GT*} applied to observed table *SGT*; see Allison, 1980; Goodman, 1971). There is no need to include three-variable interaction effects of Sex and Age on Education in the final modified LISREL model.

To find a more parsimonious model for the effects of Sex (*S*), Age (*G*), and Education (*T*) on System Involvement (*Y*), the estimated effect parameters for marginal table *SGTY* were inspected using the estimated expected frequencies of Model 1, $\hat{F}^{SGTYZABCDE}_{ijkr++++++}$ (or $\hat{F}^{SGTYZABCDE}_{ijkr+lm++p}$, with identical results, as follows from the collapsibility theorem). The sizes of the effects clearly suggested the inclusion of direct effects of Sex, Education, and Age on *Y* as well as three-variable interaction effects of Sex and Education on *Y*. However, the inclusion of the other three-variable interaction effects on *Y*, as well as the four-variable interaction effects, did not seem to be necessary.

All this leads to a modified LISREL model with estimated expected frequencies $\hat{F}^*$ that are computed in such a manner that marginal table *SGT* with frequencies $\hat{F}^{*SGTYZABCDE}_{ijk+++++++}$ satisfies model{*SG,ST,GT*}, that marginal table *SGTY* with frequencies $\hat{F}^{*SGTYZABCDE}_{ijkr++++++}$ is in agreement with model {*SGT,STY,GY*}, and that to the complete table model {*SGTYZ, AY, BY, EY, CZ, DZ*} still applies. The test statistics of this further amended modified LISREL model are presented in Table 5.2, Model 3. Model 3 does not fit the data, but it does not fit significantly worse than Models 1 or 2 (see Models 7 and 8 in Table 5.2). A model without any effects of Sex, Age, and Education on System Involvement but otherwise similar to Model 3 does fit the data much worse: $L^* = 625.218$, $df^* = 339$, $p = .000$, $\chi^2 = 615.585$.

In order to find a suitable parsimonious model for the effects of Sex (*S*), Age (*G*), Education (*T*), and System Involvement (*Y*) on Protest Tolerance (*Z*), table *SGTYZ* with frequencies $\hat{F}^{SGTYZABCDE}_{ijkrs+++++}$ or equivalently, as follows from the collapsibility theorem, table *SGTYZABCDE* with frequencies $\hat{F}^{SGTYZABCDE}_{ijkrslmnop}$) was set up using the estimated expected frequencies of Model 1. However, several entries in estimated table *SGTYZ* appeared to be empty. For example, there were no women in the age range 58-91 years old who scored low on System Involvement and high on Protest Approval. Because estimated zero cell entries make it impossible to find estimates of the loglinear parameters, zero cells were replaced by small constants (.1, .01, .001). Incredibly large parameter estimates were obtained that varied wildly for the different constants. Another approach was needed.

Instead of starting out with the saturated model for marginal table *SGTYZ*, a parsimonious logit model was defined in which there were no effects of the external and latent variables on *Z*, Protest Tolerance. Using the notation from Table 5.2, modified LISREL model {*sgt* {*SG,ST,GT*} *sgty* {*SGT,STY,GY*} total {*SGTY,AY,BY,EY,CZ,DZ*}} was set up. The test results for this parsimonious model that did not contain estimated zero cells are $L^* = 652.315$, $df^* = 357$, $p = .000$, $\chi^2 = 644.654$, the worst fit obtained so far.

The direct effects of *S, G, T,* and *Y* on *Z* were then added, one at the time, comparing each new model with the baseline model of "no effects on *Z*." Next, higher-order effects on *Z* were successively added up to the saturated model for table *SGTYZ*. Hierarchically nested models were tested against each other. The loglinear effects on *Z* were computed for each new model in which this was possible given the absence of fitted zero cells.

Age and Education turned out to have a clear direct influence on Protest Tolerance. With regard to the influence of Sex, the situation was less straightforward. For several models, especially those that contained higher-order interaction effects of Age, Sex, and System Involvement on Protest Tolerance, the estimated cell entries of table *SGTYZ* were extremely small or zero. It often was not clear whether these very small estimated expected cell frequencies were actually zeros that had not been not estimated accurately enough or were indeed just very small estimated cell frequencies. Not surprisingly, then, several closely related models yielded very different estimates of the higher-order effects on Protest Tolerance, especially those higher-order effects involving Sex. The estimates of the lower-order effects were much more stable for the various models.

Moreover, when comparing models with and without main and interaction effects of Sex on *Z*, the conditional tests yielded borderline results, *p* values around .05. An additional difficulty with these conditional tests is that when two nested models are compared, one of which involves boundary estimates for the parameters resulting from estimated zero cell frequencies, the difference between their L^2 values does not approximate the theoretical chi-square distribution. As noted above, models containing higher-order interactions of Sex often led to estimated zero cells.

Given these results and uncertainties, and in the light of the small main effect of Sex on Protest Tolerance found in most models, it was decided to give prevalence to the principle of parsimony (Hagenaars,

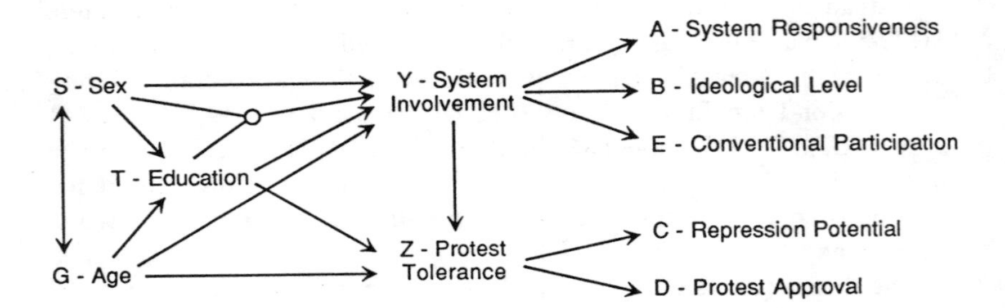

Figure 5.1. A Modified LISREL Model for Table 5.1; Model 4 in Table 5.2
NOTE: Manifest variables, *A, B, C, D, E*; latent variables, *Y, Z*; external variables, *S, T, G*.

1990, p. 61) and leave the effects of Sex on Protest Tolerance out of the model.

Another point that was not fully clear in the first instance was whether to include three-variable interaction terms for the effects of Age and System Involvement on Protest Tolerance. The outcomes of the conditional testing procedure were ambiguous. However, the estimated higher-order interaction effects concerned were very small, and it would not make any substantive difference whether they were included or not. So it was decided to leave them out, too.

Together with all other restrictions for the causally prior marginal tables and submodels, the above considerations led to the final Model 4 in Table 5.2, in which Protest Tolerance is influenced by Age, Education, and System Involvement but not by Sex, and in which no three- or more-variable interaction effects on Protest Tolerance appear. According to the test statistics, Model 4 does not fit the data, but it is certainly not worse than the other models considered.

For the time being, Model 4 is treated as our best guess about the causal relations among the variables in Table 5.1, and its outcomes are discussed in detail. The path diagram corresponding with Model 4 is presented in Figure 5.1; the left-hand side of Table 5.3 shows the parameter estimates of Model 4.

The "subtables" mentioned in Table 5.3 are marginal tables; they are formed by summing the estimated expected frequencies F^* of Model 4 over the appropriate subscripts. Marginal table *SG* is used to determine the relationship between the exogenous variables Age and Sex. The $\hat{\lambda}_{1j}^{SG}$ parameters show that women are somewhat older than men. Sub-

TABLE 5.3
Parameter Estimates of Models 4 and 5 (Table 5.2)

Effect		Model 4 $\hat{\tau}$	Model 4 $\hat{\lambda}$	Model 5 $\hat{\tau}$	Model 5 $\hat{\lambda}$	Effect	
Subtable SG						*Subtable SG*	
SG	11	1.106	.101	1.106	.101	*SG*	11
	12	1.047	.046	1.047	.046		12
	13	.863	−.147	.863	−.147		13
Subtable SGT						*Subtable SGT*	
ST	11	1.099	.095	1.099	.095	*ST*	11
GT	11	1.229	.206	1.229	.206	*GT*	11
	21	1.024	.023	1.024	.023		21
	31	.796	−.229	.796	−.229		31
Subtable SGTY						*Subtable SGTY*	
SY	11	1.553	.440	1.666	.510	*SY*	11
GY	11	.730	−.314	.815	−.204	*GY*	11
	21	1.287	.252	1.138	.130		21
	31	1.064	.062	1.077	.075		31
TY	11	2.539	.932	2.693	.990	*TY*	11
STY	111	1.391	.330	1.487	.397	*STY*	111
Subtable SGTYZ						*Subtable SGTYZABE*	
TZ	11	1.152	.142	1.098	.093	*TZ*	11
GZ	11	2.519	.925	1.606	.474	*GZ*	11
	21	.869	−.140	.990	−.010		21
	31	.457	−.785	.629	−.464		31
YZ	11	1.495	.402	1.308	.269	*YZ*	11
Effects on Indicators						*Effects on Indicators*	
Total Table						*Subtable SGTYABE*	
AY	11	.696	−.362	.693	−.367	*AY*	11
BY	11	2.261	.816	2.237	.805	*BY*	11
EY	11	.561	−.578	.567	−.568	*EY*	11
		—	—	1.183	.168	*GE*	11
		—	—	.817	−.203		21
		—	—	1.035	.034		31
						Total Table	
CZ	11	1.648	.500	1.285	.250	*CZ*	11
DZ	11	.452	−.795	.057	−2.864	*DZ*	11
		—	—	1.423	.353	*GC*	11
		—	—	.870	−.139		21
		—	—	.808	−.214		31

table *SGT* provides the direct, partial effects on Education. From parameter $\hat{\lambda}_{1\,1}^{ST}$ it follows that men are a bit better educated than women.

The parameters $\hat{\lambda}_{j\,1}^{GT}$ show that the partial odds of having had at least some college education rather than no college clearly decrease with age. Exactly the same parameter estimates were obtained above in the separate analyses in which model {*SG*} was applied to the observed table *SG* and model {*SG, ST, GT*} to the observed frequencies of table *SGT* (see Allison, 1980; Goodman, 1971).

The direct effects on System Involvement are presented under the heading "Subtable *SGTY.*" Education has very strong effects: The partial odds of having high system involvement rather than low are much larger for those who had some college than for those who had no college education ($\hat{\lambda}_{1\,1}^{TY}$) . The parameter $\hat{\lambda}_{1\,1}^{SY}$ providesthe estimate for the effect of Sex on *Y*; the system involvement of men is higher than that of women. Age has a curvilinear relationship with System Involvement, as appears from $\hat{\lambda}_{j\,1}^{GY}$: The youngest age group is the less involved, the middle group the most, and the oldest age category occupies an in-between position.

It is interesting to compare these age effects on System Involvement with the age effects found in Chapter 4. In the previous analyses, the middle age group also appeared to be the most involved, but there was no different involvement between the young and the old. The difference between the present and the previous analysis is that in the previous analysis (see Figure 4.1) age effects on System Involvement are estimated holding Sex constant, whereas in the present analysis (see Figure 5.1) age effects are determined holding Sex *and* Education constant. As it appears now, in terms of strict age effects, holding Education constant, the youngest age group is the least involved in the political system. The youngest age group compensates for this low level of involvement by having the highest education (see Table 5.3, subtable *SGT*). Because a high educational level causes high system involvement, the young appear as much involved as the old when Education is not held constant.

There is, further, an interesting three-variable interaction effect $\hat{\lambda}_{1\,1\,1}^{STY}$ of Education and Sex on System Involvement. The average main effect $\hat{\lambda}_{1\,1}^{SY}$ of Sex on *Y* is .440. According to $\hat{\lambda}_{1\,1\,1}^{STY}$, this effect is .330 larger among those with a high level of education and .330 smaller among the low-level group. Among the highly educated, men are much more involved in the political system than women; a corresponding but much smaller male-female difference exists among the less educated. Formulated the other way around, $\hat{\lambda}_{1\,1}^{TY}$ (= .932), the main effect of Education on System Involvement is .330 higher among men and .330

lower among women. The effects of education are much larger among men than among women, although in the same direction. The highly educated men are by far the most involved in the political system, more than men with a low educational level and more than women in general regardless of their level of education.

Subtable *SGTYZ* presents the direct effects on Protest Tolerance. Whereas System Involvement is clearly Education related, Protest Tolerance is an Age phenomenon: Age has very large effects on the partial odds of having a high protest tolerance instead of a low one ($\hat{\lambda}^{GZ}_{j\,1}$) . The younger one is, the more tolerant one tends to be. The effect $\hat{\lambda}^{YZ}_{1\,1}$ of Y on Z is reasonably large; the more involved in the system, the larger the protest tolerance. Finally, from $\hat{\lambda}^{YZ}_{1\,1}$, college-educated people are a bit more tolerant toward protests than are non-college-educated people.

The last effects presented in Table 5.3 are those between the latent variables and their indicators. These do not give rise to interpretations of the latent variables that are different from the ones offered so far. Also, the sizes of the effects are very much the same as we have seen before.

With these last effects between the latent variables and their indicators, all arrows in Figure 5.1 are described and their effects estimated. Table 5.3 does not contain the standard errors of the estimates. The programs that were used (and that have been available up to now) do not compute standard errors for the parameters of modified LISREL models. The significance of a particular effect has to be determined by using the conditional test statistic $L_{\mathrm{r/u}}$, comparing two models, one with and one without the effect concerned. This is a statistically sound but somewhat time-consuming procedure.

In addition to direct effects, investigators are often interested in *indirect and total effects.* In loglinear analysis these indirect and total effects cannot be obtained in a simple manner. There is no calculus for modified LISREL models comparable to that of LISREL or path models (see Chapter 2). Total loglinear effects of a particular independent variable on a particular dependent variable must be found by computing the relevant effect parameter in certain marginal tables. These marginal tables are formed by summing the estimated expected frequencies $\hat{F}^{*}$ over all variables, but not over the independent and the dependent variable in which one is interested and not over the antecedent variables that cause spurious relationships between the independent and the dependent variable concerned. For example, to obtain an estimate of the total effect of Education T on Protest Tolerance Z, one uses marginal table $SGTZ$ with frequencies $\hat{F}^{*SGTYZABCDE}_{i\,j\,k+s+++++}$ or, with identical results as follows from the collapsibility theorem, marginal table GTZ with frequencies $\hat{F}^{*SGTYZABCDE}_{+j\,k+s+++++}$. The estimated expected frequencies $\hat{F}^{*}$ are summed over the indicators (A through E are causally posterior to Z) and over Y (this latent variable comes in between T and Z), but not over G and S as these variables are sources of spurious association between T and Z and therefore should be held constant when one is determining the total effect of T on Z. Applying an appropriate model, such as the saturated model, to marginal table $SGTZ$ with frequencies $\hat{F}^{*SGTYZABCDE}_{i\,j\,k+s+++++}$, the total causal effect of Education on Protest Tolerance can be obtained. The total effect of T on Z in marginal table $SGTZ$ is estimated to be $\hat{\lambda}^{TZ}_{1\,1} = .356$ ($\hat{\tau}^{TZ}_{1\,1} = 1.426$), which is much larger than the direct effect $\hat{\lambda}^{TZ}_{1\,1} = .142$ presented in Table 5.3, Subtable $SGTYZ$. Although this procedure does not lead to an elegant decomposition of the marginal effects in terms of partial direct, indirect, and spurious effects, it does provide insight into the overall importance of a particular variable as a cause of another variable.

These conclusions about direct and total effects rest on the assumption that Model 4 is a valid description of the population. For several reasons, the test results presented do not provide firm evidence regarding whether or not this is the case.

In our search for the true population model, a number of tests have been carried out on the same data. In that situation, the significance levels cannot be given the standard interpretation, and modifications of the usual procedures are called for (Aitkin, 1980).

Using the testing procedures in an exploratory fashion, as is done here, brings out the dangers of under- and overfitting. The former refers to the possibility that effects that exist in the population are not incorporated into the final model; the latter, to the possibility that effects that do not exist in the population appear to be significant and are added to the final model. In earlier work I have summarized a number of steps recommended in the literature for avoiding these dangers (Hagenaars, 1990, pp. 63-64). Because we have milked Table 5.1 dry in our search for large significant effects that may be included and for small, nonsignificant effects that might be left out, the dangers of over- and underfitting pose a serious threat.

Ideally, one should repeat the analyses on a new sample, and preferably on a larger sample. The power of chi-square tests for the models in Table 5.2 is rather low, even with a sample size of 1,156. And what is worse, given the many empty or nearly empty estimated cell frequencies, L^2 and χ^2 cannot be expected to approximate the theoretical chi-square distribution closely. Also for this reason, the meaning of the significance levels p in Table 5.2 is unclear, and little importance should be attached to the circumstance that the final Model 4 in Table 5.2 does not fit ($p = .005$).

Nevertheless, it is possible to obtain a "well-fitting" model. Inspection of the residual frequencies of Model 4 suggested that Model 4 might be improved by adding direct connections between some external variables and some indicators. Introducing these effects yielded a well-fitting model. But most extra effects were very small, except for the relations between Age (G) and Repression Potential (C) and Age and Conventional Participation (E). If only the effects GC and GE are added to Model 4, the result is a well-fitting model whose test statistics are reported in Table 5.2, Models 5 and 12.

As is clear from the description of Model 5 in Table 5.2, because of the direct effects of Age on indicators E and C, more steps and more marginal tables are needed to obtain the estimated expected frequencies $\hat{F}^*$.

The parameter estimates of Model 5 are presented on the right-hand side of Table 5.3. Whether Model 5 may be considered an acceptable model from a substantive point of view depends on whether sensible interpretations can be found for the direct effects of Age on indicators E and C over and above the effects Age has on latent variables Y and Z. The effects G-E and G-C are mentioned under the heading "Effects on Indicators." An interpretation of these effects must probably be found in terms of "test item bias," a phenomenon mentioned in Chapter 4. In this case, the occurrence of test item bias implies that the meaning of the indicators E and C varies according to age. If the effects G-E and G-C can be meaningfully interpreted along these lines, Model 5 provides the estimates for the relationships in Figure 5.1 corrected for test item bias. If no meaningful interpretation of the effects G-E and G-C is possible, judgment on the validity of Model 5 should be suspended until it is tested on new data.

In order to find a meaningful explanation of the effects G-E and G-C, one must take a close look at the original separate questions by means of which Conventional Participation and Repression Potential have been measured. This is beyond the scope of this monograph.

Comparison of the outcomes of Models 4 and 5 in Table 5.3 indicates that the consequences of these extra age effects on indicators C and E are not very dramatic in substantive terms. Although the sizes of several effects changed, introduction of the direct effects of Age on indicators C and E affected neither the signs of the other effects nor the order of their magnitudes. The main conclusions drawn from the parameter estimates of Model 4 still hold for Model 5.

6. LATENT VARIABLE MODELS FOR LONGITUDINAL DATA

The analysis of social and behavioral change is full of pitfalls. One of the greatest dangers is the confounding of underlying true change with manifest change caused by unreliability of the measurements. Although it is defined as error, as a random phenomenon, unreliability can produce patterns of manifest change that look very systematic.

For studying change in categorical characteristics, loglinear modeling with latent variables offers an excellent framework within which the systematic consequences of unreliability may be discovered and accounted for. The essential characteristics of the latent variable approach to the analysis of longitudinal data will be shown in the next section, employing a simple 2×2 turnover table. Later on, more complicated situations are introduced. (A more complete survey is provided in Hagenaars, 1990, 1992, in press.)

Manifest Change and Latent Stability: Analyzing a 2 × 2 Turnover Table

Table 6.1 presents the observed changes in the intentions of Dutch respondents to vote or not to vote between November and December 1971. According to Table 6.1, 16.5% [= 100 × (43 + 95)/837)] of the respondents changed their intentions between November and December. The horizontal percentages indicate that especially the intentions of the nonvoters are unstable. Whereas 93.8% of those who intended to vote in November still intended to vote in December, the corresponding stability percentage for the nonvoters is merely 33.1%.

Once the possibility of measurement error is allowed, the question arises to what extent the manifest changes reflect unreliability rather than true change. The most extreme answer is that there is no true change at all and that the observed changes result solely from measurement error. In terms of latent class modeling, each and every person belongs either to the latent class "voter" or to the latent class "nonvoter" without changing her or his latent position. Manifest changes occur only because the latent positions

TABLE 6.1
Vote Intention in the Netherlands: November 1971-December 1971

	B. December 1971					
	1. Voter		*2. Nonvoter*		*Total*	
A. November 1971	%	*n*	%	*n*	%	*n*
1. Voter	93.8	(652)	6.2	(43)	100	(695)
2. Nonvoter	66.9	(95)	33.1	(47)	100	(142)
Total	89.2	(747)	10.8	(90)	100	(837)

SOURCE: Data from Hagenaars (1990, p. 184).

are not perfectly measured. If this is true, the data in Table 6.1 can be explained by a standard latent class model with two indicators *A*, the observed vote intention in November, and *B*, the observed vote intention in December, and one dichotomous latent variable *X*, the true vote intention, in which $X = 1$ refers to true voters and $X = 2$ to true nonvoters:

$$\pi_{i\,j\,t}^{ABX} = \pi_t^X \pi_{i\,t}^{\bar{A}X} \pi_{j\,t}^{\bar{B}X} . \tag{18}$$

Formulated as a loglinear model, the equivalent equation in multiplicative form is

$$F_{i\,j\,t}^{ABX} = \eta \tau_i^A \tau_j^B \tau_t^X \tau_{i\,t}^{AX} \tau_{j\,t}^{BX} . \tag{19}$$

This standard latent class model $\{AX,BX\}$ is not identifiable for Table 6.1. Equation 18 contains five unknown nonredundant parameters, and Table 6.1 provides only three independent knowns. If there is reason to believe that the "reliabilities" (i.e., the conditional probabilities of giving the "correct" answer in agreement with one's latent class) are the same for the true voters and the true nonvoters, the number of parameters can be reduced by two and the model becomes exactly identifiable. The relevant restrictions are

$$\pi_{1\,1}^{\bar{A}X} = \pi_{2\,2}^{\bar{A}X} \qquad \pi_{1\,1}^{\bar{B}X} = \pi_{2\,2}^{\bar{B}X} . \tag{20}$$

In terms of loglinear modeling the restrictions in Equation 20 are identical to setting the one-variable parameters for *A* and *B* in Equation 19 equal to 1 (Hagenaars, 1990, p. 185):

TABLE 6.2
Two-Latent-Class Model for Data on Dutch Vote Intention in Table 6.1

		A. Vote Intention November 1971 $\hat{\pi}_{it}^{\bar{A}X}$		*B. Vote Intention December 1971* $\hat{\pi}_{jt}^{\bar{B}X}$	
Latent Class X_t	$\hat{\pi}_t^X$	*1. Voter*	*2. Nonvoter*	*1. Voter*	*2. Nonvoter*
1	.940	.876	.124	.946	.054
2	.060	.124	.876	.054	.946
		$\hat{\tau}_{11}^{AX}$ = 2.653		$\hat{\tau}_{11}^{BX}$ = 4.184	
		$\hat{\lambda}_{11}^{AX}$ = 0.976		$\hat{\lambda}_{11}^{BX}$ = 1.431	
$L^2 = 0$, df = 0					

NOTE: Identical parameter estimates result from a priori equality restrictions.

$$\tau_i^A = \tau_j^B = 1. \tag{21}$$

This restricted model $\{AX, BX\}$ has zero degrees of freedom and fits the observed frequencies in Table 6.1 perfectly ($L^2 = 0$). The parameter estimates are presented in Table 6.2.

According to the manifest data in Table 6.1, the percentage of voters was 83.0% in November and 89.2% in December. In the latent class model in Table 6.2, the percentage of true voters is estimated at 94.0%. The distribution of the latent variable Vote Intention is more skewed and has smaller estimated variance $[\hat{\pi}(1 - \hat{\pi})]$ than the manifest distributions of Vote Intention. This result resembles the consequences of the postulates of classical test theory, where the observed variance equals the sum of the true variance and the error variance (Bohrnstedt, 1983, p. 71).

The observed marginal distribution of B is closer to the true distribution of Vote Intention than the observed marginal distribution of A. This is a consequence of the greater reliability of the measurements in December (B) than in November (A), as can be seen from a comparison of the estimates $\hat{\pi}_{ij}^{\bar{A}X}$ with $\hat{\pi}_{ij}^{\bar{B}X}$ or $\hat{\tau}_{11}^{AX}$ with $\hat{\tau}_{11}^{BX}$ in Table 6.2.

Restricted model $\{AX, BX\}$ provides a reasonable explanation of the manifest turnover in Vote Intention in the sense that the estimates of the latent class parameters are substantively plausible. However, it is not possible to test the basic postulate of the model that there is no true change. For the estimation of models that allow manifest change to be produced by a combination of unreliability and true latent change, more data are required: manifest variables with more than two categories, two or more indicators for each latent variable, or measurements from more than two waves. Some of these possibilities will be discussed below.

Before dealing with these more complicated models, the simple no-latent-change model for Table 6.1 may be used to point out another misleading consequence of measurement unreliability. As mentioned above, the horizontal percentages in Table 6.1 lead to the conclusion that the vote intention of the November Voters is much more stable than that of the November Nonvoters. However, assuming that the simple no-latent-change model in Table 6.2 is valid, these conclusions about differences in stability between voters and nonvoters are completely misleading. In the first place, no true changes take place at all among either the voters or the nonvoters. In the second place, as far as unreliability reflects incidental but "real" fluctuations (Hagenaars, 1990, sec. 4.4.2), these oscillations have the same sizes for voters and nonvoters because the reliabilities in the model in Table 6.2 are the same for both latent classes. As Maccoby (1956) has noted, given the assumptions that there is no true change and that the unreliability is more or less the same for all classes, the result is a manifest turnover table with equal absolute cell frequencies (1,2) and (2,1) but with horizontal percentages in which the smallest category appears to be the most unsteady.

The many studies in which it is argued that the supporters of the smaller political parties are less loyal and more easily switch their preferences than the followers of the major parties (Barnes, 1989), that the owners of cars of a major brand are more loyal than those possessing minor brand cars (Zeisel, 1968, chap. 13), and so on, all have to demonstrate that their substantive results are not a simple consequence of even small amounts of unreliability of the measurements.

Latent Changes in One Characteristic

Table 6.3 is used to illustrate loglinear models with latent variables in which manifest changes may be the result not only of measurement errors but also of the occurrence of true changes. Table 6.3 is a $2 \times 2 \times 2$ turnover table, taken from a classic American voting study conducted in 1948 by Berelson, Lazarsfeld, and McPhee (1954). Wiggins (1955,

TABLE 6.3
Vote Intention, Elmira, New York, 1948

A. June	*B. August*	*C. October* *1. Republican*	*2. Not Republican*	*Total*
1. Republican	1. Republican	307	10	317
	2. Not Republican	13	32	45
2. Not Republican	1. Republican	46	10	56
	2. Not Republican	9	135	144
	Total	375	187	562

SOURCE: Data from Wiggins (1973, p. 136).

1973) analyzed these data in the late 1940s from the viewpoint of separating true, latent change from change caused by unreliability.

Assuming perfect measurements, the observed changes in Table 6.3 can be analyzed using ordinary loglinear models without latent variables. Model $\{AB, AC, BC\}$ turns out to be the best and only choice. It fits almost perfectly: $L^2 = .018\ (= \chi^2)$, $\text{df} = 1, p = .892$. The Markovian model $\{AB, BC\}$, in which there is no direct relationship between Vote Intention in June and October, has to be rejected: $L^2 = 29.110$, $\text{df} = 2, p = .000, \chi^2 = 37.359$. (Van de Pol & Langeheine, 1990, give a full treatment of the Markov model with possibly constant transition rates, defined at the latent or the manifest level, and taking into account a heterogeneous population.)

In model $\{AB, AC, BC\}$, the direct relation Vote Intention August-October is very strong: $\hat{\lambda}_{1\ 1}^{BC} = 1.071$. The direct relation Vote Intention June-October is much smaller: $\hat{\lambda}_{1\ 1}^{AC} = .463$. The parameter for the relation Vote Intention June-August has to be estimated by means of marginal table AB according to the principles of the modified path approach (see Chapter 2), with the result $\hat{\lambda}_{1\ 1}^{AB} = .724$. Because of the longitudinal nature of the data, the reported two-variable effects may be interpreted as measures of stability. It can then be concluded that voter intentions changed more in the beginning of the election campaign than at the end.

But what if the measurements are not completely reliable? If the extreme position is adopted that no true change occurred and that all manifest change results from measurement error, the standard two-latent-class model $\{AX, BX, CX\}$ depicted in Figure 6.1a is assumed to be valid.

Model $\{AX, BX, CX\}$ fits the $2 \times 2 \times 2$ observed table perfectly, but with zero degrees of freedom. The estimates of the conditional response

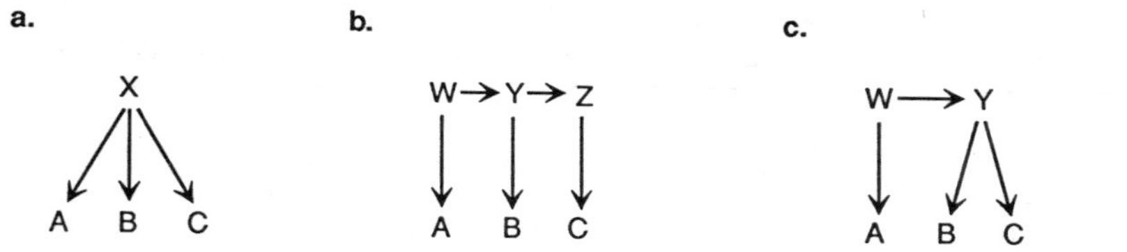

a. No Latent Change

b. 'Markovian' Latent Change

c. 'Socratic' Latent Change

Figure 6.1. Loglinear Models With Latent Variables for Three Waves
NOTE: W, X, Y, Z are latent variables; A, B, and C are measurements of the same characteristics at Waves 1, 2, and 3, respectively.

probabilities (not shown here) indicate that the true Republicans ($X = 1$) have a somewhat greater chance of stating Vote Intention correctly than do the true non-Republicans ($X = 2$) and that for both latent classes the conditional probabilities of giving the correct answer increase over time, the most between June and August.

In order to test whether the increase in reliabilities over time is significant, the following restrictions are imposed on the conditional response probabilities of model $\{AX, BX, CX\}$:

$$\pi_{1\,1}^{\bar{A}X} = \pi_{1\,1}^{\bar{B}X} = \pi_{1\,1}^{\bar{C}X} \quad \text{and} \quad \pi_{2\,2}^{\bar{A}X} = \pi_{2\,2}^{\bar{B}X} = \pi_{2\,2}^{\bar{C}X}, \tag{22}$$

which in terms of loglinear parameters is identical to the restrictions

$$\tau_{1\,1}^{AX} = \tau_{1\,1}^{BX} = \tau_{1\,1}^{CX} \quad \text{and} \quad \tau_{1}^{A} = \tau_{1}^{B} = \tau_{1}^{C}\,. \tag{23}$$

The thus restricted model $\{AX, BX, CX\}$ does not fit: $L^2 = 50.698$, df $= 4$, $p = .000$, $\chi^2 = 54.578$. The increase of the reliabilities over time is significant.

However, introducing the restriction that the reliabilities for latent classes 1 and 2 are the same, while varying over time,

$$\pi_{1\,1}^{\bar{A}X} = \pi_{2\,2}^{\bar{A}X}, \quad \pi_{1\,1}^{\bar{B}X} = \pi_{2\,2}^{\bar{B}X}, \quad \pi_{1\,1}^{\bar{C}X} = \pi_{2\,2}^{\bar{C}X}, \tag{24}$$

or, equivalently, in loglinear terms,

$$\tau_{1}^{A} = \tau_{1}^{B} = \tau_{1}^{C} = 1 \tag{25}$$

TABLE 6.4
Two-Latent-Class Model for Elmira Vote Intention Data in Table 6.3

		X. Vote Intention Class 1 Republicans	*X. Vote Intention* Class 2 Not Republicans	
A. Vote Intention June	1. Republican	.850[a]	.150	$\hat{\lambda}^{AX}_{11}$ = .868
	2. Not Republican	.150	.850	
B. Vote Intention August	1. Republican	.956	.045	$\hat{\lambda}^{BX}_{11}$ = 1.533
	2. Not Republican	.045	.956	
C. Vote Intention October	1. Republican	.966	.034	$\hat{\lambda}^{CX}_{11}$ = 1.680
	2. Not Republican	.034	.966	
	Latent class size $\hat{\pi}^X_t$	.678	.324	

$L^2 = 4.077$, df = 3, $p = .253$
$\chi^2 = 4.162$

a. Entries in this column and the next are, except for the last row, the conditional response probabilities $\hat{\pi}^{\bar{A}X}_{it}$, $\hat{\pi}^{\bar{B}X}_{jt}$, and so on; identical estimates result from a priori equality restrictions.

results in a fitting model. Table 6.4 presents the test statistics and the parameter estimates for this model. Note that the true support for the Republicans is much larger than for the non-Republicans ($\hat{\pi}^X_t$) and that the reliabilities of the measurements increase over time, in particular between June and August ($\hat{\lambda}^{AX}_{11}$, $\hat{\lambda}^{BX}_{11}$, $\hat{\lambda}^{CX}_{11}$).

Although the restricted model $\{AX, BX, CX\}$ in Table 6.4, in which the manifest change is solely attributed to measurement error, fits the data in Table 6.3 very well, there are alternative explanations of these data, which in addition allow for the possibility that real changes in vote intention took place during the 1948 Elmira election campaign. If true changes occurred at all three points in time, a three-latent-variable model is required in which latent variable W indicates the true Vote Intention in June, with indicator A; latent variable Y the true Vote Intention in August, with indicator B; and latent variable Z the true Vote Intention in October, with indicator C. All three latent variables are regarded as dichotomies with categories 1, Republican, and 2, not Republican.

Without further restrictions, model $\{WYZ, AW, BY, CZ\}$ is not identifiable because it has too many unknowns. Restrictions may be found by assuming Markovian latent change. In spite of the evidence from the manifest table, Model $\{WY, YZ\}$ is postulated for the relations among the latent variables (Figure 6.1b). In addition, it may be assumed that the conditional response probabilities of giving the correct answer are stable over time and are the same for the true Republicans and the true non-Republicans. Besides the restrictions on the conditional response probabilities $\pi_{i\,q\,r\,s}^{\bar{A}WYZ}$ that are needed to ensure that they vary only with categories of W, and not with Y or Z, and analogous restrictions for $\pi_{j\,q\,r\,s}^{\bar{B}WYZ}$ and $\pi_{k\,q\,r\,s}^{\bar{C}WYZ}$ (see Table 3.3), the following extra restrictions are imposed:

$$\pi_{1\,1\,r\,s}^{\bar{A}WYZ} = \pi_{2\,2\,r\,s}^{\bar{A}WYZ} = \pi_{1\,q\,1\,s}^{\bar{B}WYZ} = \pi_{2\,q\,2\,s}^{\bar{B}WYZ} = \pi_{1\,q\,r\,1}^{\bar{C}WYZ} = \pi_{2\,q\,r\,2}^{\bar{C}WYZ}, \tag{26}$$

or, in loglinear terms,

$$\tau_{1\,1}^{AW} = \tau_{1\,1}^{BY} = \tau_{1\,1}^{CZ} \quad \text{and} \quad \tau_{1}^{A} = \tau_{1}^{B} = \tau_{1}^{C} = 1. \tag{27}$$

Model $\{WY, YZ, AW, BY, CZ\}$ with the restrictions in Equation 27 fits the data well: $L^2 = 1.871$, df = 1, $p = .171$, $\chi^2 = 1.972$. According to this model, the true changes are "Markovian." Looking at the parameter estimates for the relations among the latent variables, the relation between Y and Z, the true Vote Intentions in August and October, appeared to be extremely strong: $\hat{\lambda}_{1\,1}^{YZ} = 5.473$, indicating an almost perfect association between the true Vote Intentions of August and October and, consequently, almost no latent change between these two months. In comparison, the association between the true Vote Intentions of June and August is weak: $\hat{\lambda}_{1\,1}^{WY} = .953$.

This suggests testing a two-latent-variable model $\{WY, AW, BY, CY\}$ in which the dichotomous latent variable W, with indicator A, denotes the true Vote Intention in June and the dichotomous latent variable Y, with indicators B and C, the stable true Vote Intention in August and October. Figure 6.1c shows the path diagram and Table 6.5 the test results and the parameter estimates for model $\{WY, AW, BY, CY\}$ subjected to additional restrictions of equal reliabilities for both latent classes at each point in time, analogous to Equations 26 and 27.

The restricted two-latent-variable model $\{WY, AW, BY, CY\}$ fits the data excellently. Manifest variables A, B, and C are very reliable indicators of the latent variables. The entries of the latent turnover table WY show that from June to August an estimated 12% of the people changed their vote intentions, resulting in a net shift toward the Republican party of almost 2%.

TABLE 6.5
Two-Latent-Variable Model for Elmira Vote Intention Data in Table 6.3

		Class 1	*Class 2*	*Class 3*	*Class 4*	
W. Vote Intention June		*1. Rep.*	*1. Rep.*	*2. Not Rep.*	*2. Not Rep.*	
Y. Vote Intention August/October		*1. Rep.*	*2. Not Rep.*	*1. Rep.*	*2. Not Rep.*	$\hat{\lambda}$[a]
A. Vote Intention June	1. Republican	.961[b]	.961	.039	.039	*AW*: 1.604
	2. Not Republican	.039	.039	.961	.961	
B. Vote Intention August	1. Republican	.961	.039	.961	.039	*BY*: 1.604
	2. Not Republican	.039	.961	.039	.961	
C. Vote Intention August	1. Republican	.961	.039	.961	.039	*CY*: 1.604
	2. Not Republican	.039	.961	.039	.961	
Latent class size $\hat{\pi}^{WY}_{rs}$		.605	.051	.069	.274	WY: .961

$L^2 = 2.235$, df = 3, $p = .525$
$\chi^2 = 2.244$

a. Entries in this column are the values of the loglinear two-variable parameter $\hat{\lambda}^{AW}_{11}$, $\hat{\lambda}^{BY}_{11}$, and so on.
b. Entries in this column and the next three are, except for the last row, the conditional response probabilities $\hat{\pi}^{\bar{A}WY}_{irs}$, $\hat{\pi}^{\bar{B}WY}_{jrs}$, and so on; identical values are the result of a priori equality restrictions.

The model in Table 6.5 has been labeled "Socratic Change" in Figure 6.1c, a term borrowed from Jagodzinski, Kühnel, and Schmidt (1987). The reason for the perfect stability of the true Vote Intention between August and October might be the ongoing election campaign, in which people make up their minds more and more definitely as the campaign evolves. But it may also be an artifact of the panel study itself. As Socrates made the citizens of Athens conscious of their beliefs and opinions by questioning them, modern questionnaires may make respondents aware of their implicitly held opinions and attitudes. Because change is most likely to occur when opinions and attitudes remain implicit and less likely when an attitude or opinion has been consciously formed, change occurs more likely between the first and the second waves and not between later waves.

The pattern of manifest change in Table 6.3 can be accounted for by all three models in Figure 6.1. First, there is the Socratic change model, in which the reliabilities are stable over time but in which latent changes occur during the first waves but not in later waves (Figure 6.1c).

TABLE 6.6
Party Preference and Candidate Preference, Erie County, Ohio, 1940

	C. Party Preference-t_2 / *D. Candidate Preference-t_2*	*1. Dem.* *1. Against*	*1. Dem.* *2. For*	*2. Rep.* *1. Against*	*2. Rep.* *2. For*	*Total*
A. Party Preference-t_1	*B. Candidate Preference-t_1*					
1. Democrats	1. Against Willkie[a]	68	2	1	1	72
	2. For Willkie	11	12	0	1	24
2. Republicans	1. Against Willkie	1	0	23	11	35
	2. For Willkie	2	1	3	129	135
	Total	82	15	27	142	266

SOURCE: Data from Lazarsfeld (1972, p. 392).
a. Willkie was the Republican candidate.

Second, there is the Markovian change model (Figure 6.1b), with stable reliabilities and estimated increasing latent stability over time. Finally, there is the model in which no latent change occurs but in which the reliabilities of the indicators increase over time (Figure 6.1a). With these data it is impossible to tell on empirical grounds which explanation is correct. Theoretical considerations should decide. With an enlarged data set with more indicators per latent variable, identifiable models can be defined in which both latent change and changing reliabilities may be introduced.

Investigating Changes in Two Related Characteristics

When two or more related characteristics are measured on two or more occasions, complex manifest tables arise. Loglinear models with latent variables often provide surprisingly simple models for the complicated relations in the manifest data. Table 6.6, which was taken from another classic American voting study conducted in 1940 by Lazarsfeld, Berelson, and Gaudet (1948), is used to illustrate this point. The data concern the answers at two points in time to questions regarding whether the respondents intended to vote for the Democrats or the Republicans and whether they were against or for the Republican candidate Willkie.

Lazarsfeld used this data set in a manuscript from 1946 (published in 1972) to investigate the problem of whether the preference for a particular party caused the respondent to prefer that party's candidate or whether the preference for a particular candidate made the person join the candidate's party. Lazarsfeld's strategy essentially amounts to inspecting the changes among those who had inconsistent preferences in the first wave (i.e., those who preferred the Democratic party but at the same time the Republican candidate and those who intended to vote for the Republicans but were against the Republican candidate). The index of relative strength he developed reflects whether the inconsistencies were more often solved in Wave 2 by adapting party preference to candidate preference or by adjusting candidate preference to party preference. Party Preference appeared to be the cause of Candidate Preference, rather than the other way around. However, this conclusion is based on the assumptions that Party Preference and Candidate Preference measure distinct political preferences and that the observed turnover in Table 6.6 accurately reflects underlying true changes. (For a survey of these and other assumptions of this "cross-lagged panel correlation technique," see Hagenaars, 1990, sec. 5.4.)

Perhaps Candidate Preference and Party Preference are indicators of one underlying variable, Political Preference, that is stable over time. If these latter suppositions are true, the standard latent class model $\{AX, BX, CX, DX\}$ with one dichotomous latent variable X should be valid in the population. Figure 6.2a shows the path diagram for this model, along with the test statistics for the data in Table 6.6. Model $\{AX, BX, CX, DX\}$ must be rejected.

Relaxing the restriction that there is no latent change, but adhering to the restriction that Candidate Preference and Party Preference are both indicators of one underlying concept, Political Preference, leads to model $\{YZ, AY, BY, CZ, DZ\}$. In this model Y and Z are the underlying dichotomous variables Political Preference at the first and the second interviews, respectively. As follows from the test outcomes in Figure 6.2b, this is not a viable model either.

So perhaps Candidate Preference and Party Preference are not both indicators of one theoretical variable, but refer to distinct concepts. Allowing for this possibility but returning to the assumption of no latent change results in model $\{YZ, AY, CY, BZ, DZ\}$, in which Y and Z are dichotomous latent variables referring to the stable characteristics Party Preference and Candidate Preference, respectively. According to L^2 shown in Figure 6.2c, this is an acceptable model (although χ^2 deviates from L^2 and has a p value of $p = .021$).

We may conclude that Party Preference and Candidate Preference are not indicators of one underlying variable, but refer to distinct concepts.

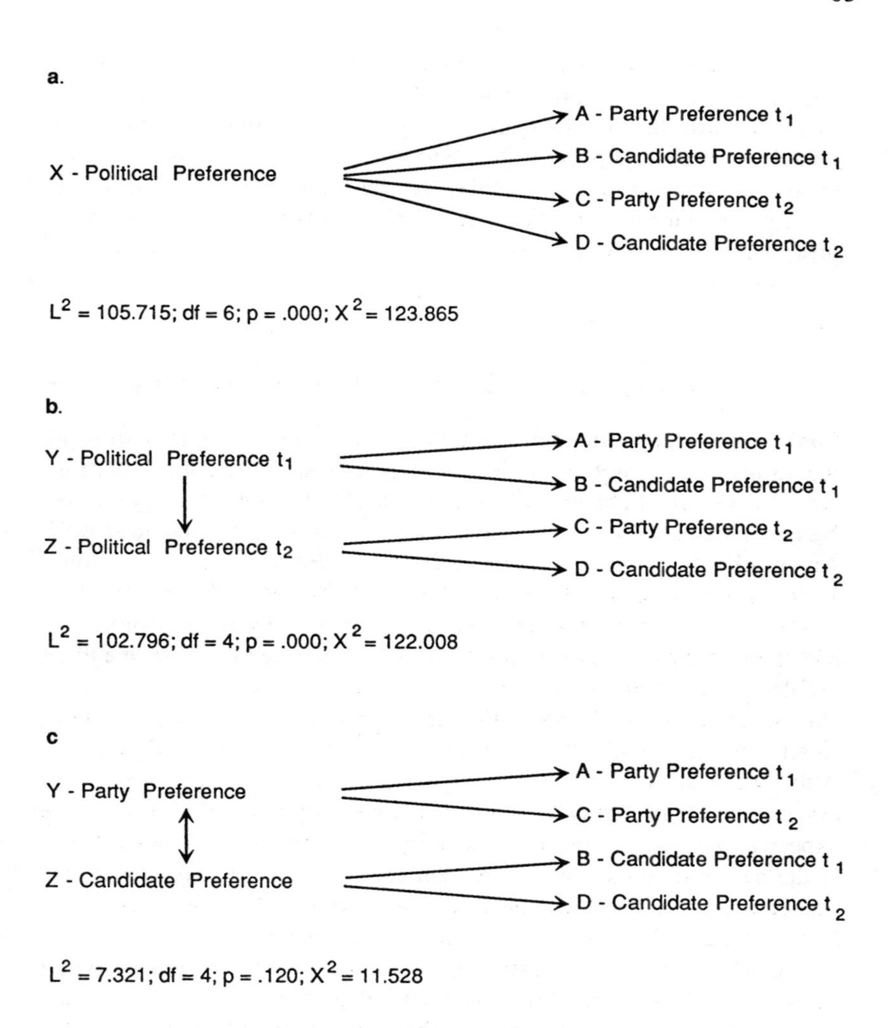

Figure 6.2. Latent Variable Models for Table 6.6

Furthermore, there is no need to assume that true changes have occurred from the first to the second wave in either characteristic. If this model is true, there is no reason to accept Lazarsfeld's conclusions about Party Preference determining Candidate Preference rather than the other way around.

The models in this section are mainly measurement models involving only latent and manifest variables. However, external variables can easily be introduced. The causal analysis of longitudinal data has its own difficulties and peculiarities, such as test-retest effects among the measurements (Hagenaars, 1988a). At the same time, the solutions to these particular problems are derived directly from the general principles of loglinear causal analysis with latent variables explained in the previous chapters.

7. PROBLEMS AND NEW DEVELOPMENTS

Loglinear models with latent variables are excellent tools for exploring the validity of theoretical notions and hypotheses concerning relations among categorical characteristics. Many highlights have been presented. Nevertheless, several shortcomings and problems have come up as well. An excellent review of *Common Problems/Proper Solutions* (Long, 1988) is presented by Clogg and Eliason (1987, also in Long, 1988).

The most important problems have been caused by small sample size and sparse tables. Many solutions have been proposed to overcome these problems, including, among others, using modifications of the standard chi-square statistics and smoothing the observed data by the addition of small constants to all cell frequencies or by more complicated methods. Although it seems hopeless to expect a solution to work well in all circumstances, in particular situations some proposed solutions may be expected to be successful. It is wise to try out several reasonable solutions. If the substantive conclusions remain the same for different solutions, accept them. If not, suspend judgment. Agresti's (1990) work is an excellent source on what to do when confronted with sparse tables.

Recently, other problems, such as item nonresponse, complex sampling designs, and the incorporation of ordinal and interval level data, have been attacked more successfully. Some of these are addressed below.

A substantial proportion of respondents is often not included in analyses because they happen to have missing scores on just one or two variables. Because small frequencies pose a problem in loglinear analysis, and item nonresponse may be selective, it is better to use all available information, even the partial information provided by the nonrespondents. Haberman (1988; Winship & Mare, 1989), Fay (1986), and Little and Rubin (1987), among others, have developed very elegant and practical procedures to estimate the parameters of loglinear models taking all available information into account. Little and Rubin (1987) provide the most comprehensive survey (also see Hagenaars, 1990, sec. 5.5.1, for a brief practical review).

Standard maximum likelihood estimation in loglinear modeling is based on the Poisson, the multinomial or the product-multinomial sampling distribution. Practically speaking, simple random sampling or stratified sampling with simple random sampling within strata is required. Many samples, however, result from more complex sampling designs involving two- or three-stage procedures, often in combination with clustering. Rao and Thomas (1988) and Lee, Forthofer, and Lorimor (1989) describe practical procedures for taking the more complex design effects into account.

All variables used in the examples in this monograph were dichotomous or trichotomous, but in any case have been treated as nominal-level variables. Mainly owing to the work of Haberman, the inclusion of categorical interval-level variables in which scores are assigned to the categories and that have (curvi)linear relationships with other variables poses no special problems (Agresti, 1990; Haberman, 1978, 1979; Hagenaars, 1990; Heinen, 1993). The interval-level variables may be external, latent, or manifest variables. Recently developed software (see Appendix A) enables the user to estimate a very general class of modified LISREL models with interval-level variables. The inclusion of strictly ordinal-level variables is still difficult, although promising work has been undertaken by, among others, Agresti, Chuang, and Kezouh (1987) and Croon (1990). As work progresses in this area, the employment of loglinear models with latent variables containing a mixture of nominal-, ordinal-, and interval-level variables may become a daily routine in the near future.

APPENDIX A: COMPUTER PROGRAMS

Programs for estimating loglinear models with latent variables are not yet part of packages such as BMDP, SAS, and SPSS-X. However, stand-alone programs, most of which are rather easy to use, are readily available.

Haberman's (1979) program LAT and its successor, NEWTON (Haberman, 1988), are completely formulated in loglinear terms (rather than in terms of Lazarsfeld's parameterization using conditional response probabilities. All kinds of restrictions, including equality and curvi(linear) restrictions, may be imposed on the loglinear effects by means of the appropriate design matrix (Evers &

Namboodiri, 1978). In that way, nominal-level and interval-level variables may be employed. With NEWTON, partial missing data can also be taken into account (Winship & Mare, 1989). All the loglinear models with latent variables discussed here can be estimated by means of LAT and NEWTON, except modified LISREL models. Although it is not impossible (see Winship & Mare, 1989, app.), there is no easy and general routine for doing this. As LAT and NEWTON make use of the scoring algorithm and a variant of the Newton/Raphson algorithm, respectively, initial parameter estimates have to be rather close to the final estimates. Although much better than LAT, even with NEWTON one still encounters difficulties in finding the appropriate initial estimates for some models and data sets. NEWTON is especially suited for the advanced user.

Clogg's (1981b) program MLLSA is a latent class analysis program written in terms of conditional response probabilities. It is now part of Eliason's CDAS package (Department of Sociology, Pennsylvania State University). It uses Goodman's (1974a, 1974b) version of the EM-algorithm described in Appendix B. All variables, observed and latent, are treated as nominal-level variables. Finding appropriate initial parameter estimates is no problem. With MLLSA, the parameters of all loglinear models with latent variables can be estimated, provided that the relations among all external and latent variables satisfy the saturated model.

A useful set of programs has been developed at the Methodology Department/WORC, Faculty of Social Sciences, Tilburg University. My own program LCAG is based on the same Goodman algorithm as MLLSA, but with the additional possibility of imposing unsaturated hierarchical modified path models on the latent level (see Hagenaars & Luijkx, 1987). By defining external variables as quasi-latent variables, one can estimate modified LISREL models. All variables are treated as nominal-level variables. Partial nonresponse can also be reckoned with. All models with latent variables discussed in this monograph were estimated by means of LCAG.

More recently, Ruud Luijkx, Willem Spoeltman, and I have developed LOGLAT. The algorithm in LOGLAT is directly based on the EM-algorithm described in Appendix B. It is written in loglinear terms and estimates the parameters of all latent variable models that are hierarchical loglinear models, including hierarchical modified LISREL models. All variables are treated as nominal-level variables. Provisions have been made to include partial nonresponse. LOGLAT is very user friendly. It can be obtained from Iec ProGamma at the University of Groningen, the Netherlands.

Ton Heinen, Marcel Croon, and Paul Vermaseren wrote DILTRAN, a program for estimating discrete latent trait models in which linear relationships exist between the categorical latent variable and nominal- or interval-level indicators. A large number of item response models can be fitted by means of this program (Heinen, 1993).

Although not yet thoroughly tested, the program LEM, written by Jeroen Vermunt, combines all attractive features of these latter programs. Moreover, it offers the extra possibilities of carrying out loglinear event history analysis and fitting logbilinear (association) models (Agresti, 1990, secs. 8.5, 8.6) for the relations among the manifest and latent variables. When tested, it may well be found to be the most flexible program available for analyzing categorical data within the loglinear framework.

APPENDIX B: THE EM-ALGORITHM

Thorough and general expositions of the EM-algorithm are provided by Dempster, Laird, and Rubin (1977) and Little and Rubin (1987). The important features of the variant of the EM-algorithm described here are presented by Haberman (1979, sec. 10.2) and Goodman (1973).

The iterative proportional fitting (IPF) procedure can be used to find the estimated expected frequencies $\hat{F}$ of hierarchical loglinear models without latent variables (Fienberg, 1980, sec. 3.3; Hagenaars, 1990, sec. 2.4). In IPF, the initial estimates of $\hat{F}$ are iteratively adapted to the observed marginal frequencies f to be reproduced by the hierarchical model.

For hierarchical loglinear models with latent variables, the complete table including the latent variables is not an observed table and the EM-algorithm has to be used. The EM-algorithm essentially consists of a repetition of two steps, the E-step and the M-step. These steps will be exemplified using an observed table *SEABC*, in which *S* (Sex) and *E* (Education) are external variables and *A, B,* and *C* indicators of latent variable *Y*. We want to fit model {*SE,SY,EY,YA, YB,YC*} to table *SEABC*.

First, initial estimates $\hat{F}^{SEYABC}_{ijsklm}$ (0) have to found that are in agreement with the assumed model. These can be obtained by roughly estimating the parameters of the model.

Second, in the E-step the estimated observed frequencies $\hat{f}^{SEYABC}_{ijsklm}$ are calculated:

$$\hat{f}^{SEYABC}_{ijsklm} = f^{SEABC}_{ijklm}(\hat{F}^{SEYABC}_{ijsklm}/\hat{F}^{SEYABC}_{ij+klm}) = f^{SEABC}_{ijklm}\pi^{SE\bar{Y}ABC}_{ijsklm}.$$

The estimated observed frequencies $\hat{f}$ are identical to the observed frequencies f when summed over the latent variable. If partial nonresponse has to be included, the necessary formula can be found in Hagenaars (1990, pp. 255-263).

In the M-step the estimated expected frequencies $\hat{F}$ obtained so far $-\hat{F}(0)$ at the first M-step are improved by means of IPF, treating the estimated observed frequencies $\hat{f}$ as if they were regular observed frequencies f. In this example, for model {*SE,SY,EY,YA,YB,YC*}, $\hat{F}^{SEYABC}_{ijsklm}$ is successively adjusted to the estimated observed marginal frequencies:

$$\hat{f}^{SEYABC}_{ij++++},\ \hat{f}^{SEYABC}_{i+s+++},\ \hat{f}^{SEYABC}_{+js+++},\ \hat{f}^{SEYABC}_{++sk++},\ \hat{f}^{SEYABC}_{++s+l+},\ \text{and}\ \hat{f}^{SEYABC}_{++s++m}.$$

It is not necessary to repeat the adjustments during the M-step until the outcomes converge. One iteration suffices (Little & Rubin, 1987, sec. 7.4). If a modified path or LISREL model is defined, and a stepwise estimation procedure is needed during the M-step, the appropriate marginal tables $\hat{f}$ and $\hat{F}$ are set up, and the postulated submodels are fitted to the estimated "observed" marginal tables. The estimated expected frequencies $\hat{F}$ obtained for each submodel are combined to obtain the estimated expected frequencies $\hat{F}^*$ for the whole modified LISREL model in the manner of Equation 11, in Chapter 2.

The estimated expected frequencies that come out of the M-step are used in the E-step to get new and better estimates of the estimated observed frequencies $\hat{f}$, which in turn are used in the M-step to improve the estimates $\hat{F}$, and so on, until the outcomes converge.

REFERENCES

AGRESTI, A. (1990) Categorical Data Analysis. New York: John Wiley.

AGRESTI, A., CHUANG, C., and KEZOUH, A. (1987) "Order-restricted score parameters in association models for contingency tables." Journal of the American Statistical Association 82: 619-623.

AITKIN, M. (1980) "A note on the selection of loglinear models." Biometrics 36: 173-178.

ALBA, R. D. (1987) "Interpreting the parameters of loglinear models." Sociological Methods and Research 16: 45-77.

ALLISON, P. D. (1980) "Analyzing collapsed tables without actually collapsing." American Sociological Review 45: 123-130.

ANDERSEN, E. B. (1990) The Statistical Analysis of Categorical Data. Berlin: Springer.

ASHER, H. B. (1976) Causal Modeling. Sage University Paper series on Quantitative Applications in the Social Sciences, 07-003. Beverly Hills, CA: Sage.

BARNES, S. H. (1989) "Partisanship and electoral behavior," in M. K. Jennings and J. W. Van Deth (eds.) Continuities in Political Action: A Longitudinal Study of Political Orientations in Three Western Democracies. Berlin: DeGruyter.

BARNES, S. H., and KAASE, M. (eds.) (1979) Political Action: Mass Participation in Five Western Democracies. Beverly Hills, CA: Sage.

BERELSON, B. R., LAZARSFELD, P. F., and McPHEE, W. N. (1954) Voting. Chicago: University of Chicago Press.

BISHOP, Y. M. M., FIENBERG, S. E., and HOLLAND, P. W. (1975) Discrete Multivariate Analysis: Theory and Practice. Cambridge: MIT Press.

BOHRNSTEDT, G. W. (1983) "Measurement," in P. H. Rossi, J. D. Wright, and A. B. Anderson (eds.) Handbook of Survey Research. Orlando, FL: Academic Press.

BOLLEN, K. A. (1989) Structural Equations With Latent Variables. New York: John Wiley.

CLOGG, C. C. (1979) Measuring Underemployment: Demographic Indicators for the United States. New York: Academic Press.

CLOGG, C. C. (1981a) "Latent structure models of mobility." American Journal of Sociology 86: 836-868.

CLOGG, C. C. (1981b) "New developments in latent structure analysis," in D. J. Jackson and E. F. Borgatta (eds.) Factor Analysis and Measurement in Sociological Research. Beverly Hills, CA: Sage.

CLOGG, C. C., and ELIASON, S. R. (1987) "Some common problems in loglinear analysis." Sociological Methods and Research 16: 8-44.

CROON, M. A. (1990) "Latent class analysis with ordered latent classes." British Journal of Mathematical and Statistical Psychology 43: 171-192.

DAVIS, J. A. (1986) The Logic of Causal Order. Sage University Paper series on Quantitative Applications in the Social Sciences, 07-055. Beverly Hills, CA: Sage.

DE LEEUW, J., VAN DER HEIJDEN, P. G., and VERBOON, P. (1990) "A latent time-budget model." Statistica Neerlandica 44: 1-22.

DeMARIS, R. (1992) Logit Modeling: Practical Applications. Sage University Paper series on Quantitative Applications in the Social Sciences, 07-086. Newbury Park, CA: Sage.

DEMPSTER, A. P., LAIRD, N. M., and RUBIN, D. B. (1977) "Maximum likelihood from incomplete data via the EM algorithm." Journal of the Royal Statistical Society, Series B, 39: 1-38.

ERIKSON, R., and GOLDTHORPE, J. H. (1992) The Constant Flux: A Study of Class Mobility in Industrial Nations. Oxford: Oxford University Press.

EVERS, M., and NAMBOODIRI, N. K. (1978) "On the design matrix strategy in the analysis of categorical data," in K. F. Schuessler (ed.) Sociological Methodology 1979. San Francisco: Jossey-Bass.

FAY, R. E. (1986) "Causal models for patterns of nonresponse." Journal of the American Statistical Association 81: 354-365.

FIENBERG, S. E. (1980) The Analysis of Cross-Classified Categorical Data. Cambridge: MIT Press.

FORMANN, A. K. (1985) "Constrained latent class analysis: Theory and applications." British Journal of Mathematical and Statistical Psychology 38: 87-111.

GOODMAN, L. A. (1971) "Partitioning of chi-square, analysis of marginal contingency tables, and estimation of expected frequencies in multidimensional contingency tables." Journal of the American Statistical Association 66: 339-344.

GOODMAN, L. A. (1972) "A modified multiple regression approach to the analysis of dichotomous variables." American Sociological Review 37: 28-46.

GOODMAN, L. A. (1973) "The analysis of multidimensional contingency tables when some variables are posterior to others: A modified path analysis approach." Biometrika 60: 179-192.

GOODMAN, L. A. (1974a) "The analysis of systems of qualitative variables when some of the variables are unobservable: A modified latent structure approach." American Journal of Sociology 79: 1179-1259.

GOODMAN, L. A. (1974b) "Exploratory latent structure analysis using both identifiable and unidentifiable models." Biometrika 61: 215-231.

HABERMAN, S. J. (1978) Analysis of Qualitative Data: Introductory Topics (Vol. 1). New York: Academic Press.

HABERMAN, S. J. (1979) Analysis of Qualitative Data: New Developments (Vol. 2). New York: Academic Press.

HABERMAN, S. J. (1988) "A stabilized Newton-Raphson algorithm for loglinear models for frequency tables derived by indirect observation," in C. C. Clogg (ed.) Sociological Methodology 1988 (Vol. 18). Washington, DC: American Sociological Association.

HAGENAARS, J. A. (1988a) "Latent structure models with direct effects between indicators: Local dependence models." Sociological Methods and Research 16: 379-405.

HAGENAARS, J. A. (1988b) "LCAG—loglinear modelling with latent variables: A modified LISREL approach," in W. E. Saris and I. N. Gallhofer (eds.) Sociometric Research: Data Analysis (Vol. 2). London: Macmillan.

HAGENAARS, J. A. (1990) Categorical Longitudinal Data: Loglinear Panel, Trend, and Cohort Analysis. Newbury Park, CA: Sage.

HAGENAARS, J. A. (1992) "Exemplifying longitudinal loglinear analysis with latent variables," in P. G. M. Van der Heijden, W. Jansen, B. Francis, and G. U. H. Seeber (eds.) Statistical Modelling. Amsterdam: Elsevier.

HAGENAARS, J. A. (in press) "Latent variables in loglinear models of repeated observations," in A. von Eye and C. C. Clogg (eds.) Latent Variables Analysis in Developmental Research. Thousand Oaks, CA: Sage.

HAGENAARS, J. A., and HALMAN, L. C. (1989) "Searching for ideal types: The potentialities of latent class analysis." European Sociological Review 5: 81-96.

HAGENAARS, J. A., and LUIJKX, R. (1987). LCAG: Latent Class Analysis Models and Other Loglinear Models With Latent Variables: Manual LCAG (Working Paper Series 17). Tilburg, Netherlands: Tilburg University, Department of Sociology.

HARMAN, H. H. (1976) Modern Factor Analysis. Chicago: University of Chicago Press.

HAYS, W. L. (1981) Statistics (3rd ed.). New York: Holt, Rinehart & Winston.

HEINEN, A. G. (1993) Discrete Latent Variable Models. Tilburg: Tilburg University Press.

HOUT, M. (1989) Following in Father's Footsteps: Social Mobility in Ireland. Cambridge, MA: Harvard University Press.

JAGODZINSKI, W., KÜHNEL, S. M., and SCHMIDT, P. (1987) "Is there a 'Socratic effect' in nonexperimental panel studies." Sociological Methods and Research 15: 259-302.

JENNINGS, M. K., and VAN DETH, J. W. (eds.) (1989) Continuities in Political Action: A Longitudinal Study of Political Orientations in Three Western Democracies. Berlin: DeGruyter.

KAUFMAN, R. L., and SCHERVISH, P. G. (1986) "Using adjusted cross-tabulations to interpret loglinear relationships." American Sociological Review 51: 717-733.

KAUFMAN, R. L., and SCHERVISH, P. G. (1987) "Variations on a theme: More uses of odds ratios to interpret loglinear parameters." Sociological Methods and Research 16: 218-255.

KIIVERI H., and SPEED, T. P. (1982) "Structural analysis of multivariate data: A review," in S. Leinhardt (ed.) Sociological Methodology 1982. San Francisco: Jossey-Bass.

KIM, J. O., and MUELLER, C. W. (1978a) Factor Analysis: What It Is and How to Do It. Sage University Paper series on Quantitative Applications in the Social Sciences, 07-013. Beverly Hills, CA: Sage.

KIM, J. O., and MUELLER, C. W. (1978b) Factor Analysis: Statistical Methods and Practical Issues. Sage University Paper series on Quantitative Applications in the Social Sciences, 07-014. Beverly Hills, CA: Sage.

KNOKE, D., and BURKE, P. J. (1980) Log-Linear Models. Sage University Paper series on Quantitative Applications in the Social Sciences, 07-020. Beverly Hills, CA: Sage.

LANGEHEINE, R., and ROST, J. (eds.) (1988). Latent Trait and Latent Class Models. New York: Plenum.

LAZARSFELD, P. F. (1950a) "The interpretation and mathematical foundation of latent structure analysis," in S. Stouffer (ed.) Measurement and Prediction. Princeton, NJ: Princeton University Press.

LAZARSFELD, P. F. (1950b) "The logical and mathematical foundation of latent structure analysis," in S. Stouffer (ed.) Measurement and Prediction. Princeton, NJ: Princeton University Press.

LAZARSFELD, P. F. (1972) "Mutual effects of statistical variables," in P. F. Lazarsfeld, A. K. Pasanella, and M. Rosenberg (eds.) Continuities in the Language of Social Research. New York: Free Press.

LAZARSFELD, P. F., BERELSON, B. R., and GAUDET, H. (1948) The People's Choice. New York: Columbia University Press.

LAZARSFELD, P. F., and HENRY, N. W. (1968) Latent Structure Analysis. Boston: Houghton Mifflin.

LEE, E. S., FORTHOFER, R. N., and LORIMOR, R. J. (1989) Analyzing Complex Survey Data. Sage University Paper series on Quantitative Applications in the Social Sciences, 07-071. Newbury Park, CA: Sage.

LEWIS-BECK, M. S. (1980) Applied Regression: An Introduction. Sage University Paper series on Quantitative Applications in the Social Sciences, 07-022. Beverly Hills, CA: Sage.

LITTLE, R. J., and RUBIN, D. J. (1987) Statistical Analysis With Missing Data. New York: John Wiley.

LONG, J. S. (1983a) Confirmatory Factor Analysis. Sage University Paper series on Quantitative Applications in the Social Sciences, 07-033. Beverly Hills, CA: Sage.

LONG, J. S. (1983b) Covariance Structure Models: An Introduction to LISREL. Sage University Paper series on Quantitative Applications in the Social Sciences, 07-034. Beverly Hills, CA: Sage.

LONG, J. S. (1984) "Estimable functions in loglinear models." Sociological Methods and Research 12: 399-432.

LONG, J. S. (ed.) (1988) Common Problems/Proper Solutions: Avoiding Error in Quantitative Research. Newbury Park: Sage.

MACCOBY, E. E. (1956) "Pitfalls in the analysis of panel data: A research note on some technical aspects of VOTING." American Journal of Sociology 61: 359-362.

MARE, R. D., and WINSHIP, C. (1991) "Loglinear models for reciprocal and other simultaneous effects," in P. V. Marsden (ed.) Sociological Methodology 1991. Oxford: Basil Blackwell.

McCUTCHEON, A. L. (1987) Latent Class Analysis. Sage University Paper series on Quantitative Applications in the Social Sciences, 07-064. Newbury Park, CA: Sage.

MOOYAART, A., and VAN DER HEIJDEN, P. G. M. (1992) "The EM-algorithm for latent class analysis with equality constraints." Psychometrika 57: 261-269.

OSTERLIND, S. J. (1983) Test Item Bias. Sage University Paper series on Quantitative Applications in the Social Sciences, 07-030. Beverly Hills, CA: Sage.

RAO, J. N. K., and THOMAS, D. R. (1988) "The analysis of cross-classified categorical data from complex sample surveys," in C. C. Clogg (ed.) Sociological Methodology 1988 (Vol. 18). Washington, DC: American Sociological Association.

REYNOLDS, H. T. (1977) The Analysis of Cross-Classifications. New York: Free Press.

SARIS, W. E., DE PIJPER, W. M., and MULDER, J. (1978) "Optimal procedures for estimation of factor scores." Sociological Methods and Research 7: 85-106.

SOBEL, M. E., HOUT, M., and DUNCAN, O. D. (1985) "Exchange, structure, and symmetry in occupational mobility." American Journal of Sociology 91: 359-372.

STEIGER, J. H. (1979a) "Factor indeterminacy in the 1930's and the 1970's: Some interesting parallels." Psychometrika 44: 157-167.

STEIGER, J. H. (1979b) "The relationship between external variables and common factors." Psychometrika 44: 93-97.

VAN DE POL, F., and LANGEHEINE, R. (1990) "Mixed Markov latent class models," in C. C. Clogg (ed.) Sociological Methodology 1990 (Vol. 20). Oxford: Basil Blackwell.

VAN DER HEIJDEN, P. G. M., MOOYAART, A., and DE LEEUW, J. (1992) "Constrained latent budget analysis," in P. V. Marsden (ed.) Sociological Methodology 1992 (Vol. 22). Oxford: Basil Blackwell.

WEISBERG, H. F. (1992) Central Tendency and Variability. Sage University Paper series on Quantitative Applications in the Social Sciences, 07-083. Newbury Park, CA: Sage.

WHITTAKER, J. (1990) Graphical Models in Applied Multivariate Statistics. Chichester: John Wiley.

WICKENS, T. D. (1989) Multiway Contingency Tables Analysis for the Social Sciences. Hillsdale, NJ: Lawrence Erlbaum.

WIGGINS, L. M. (1955) "Mathematical models for the analysis of multi-wave panels." *Doctoral Dissertation Series 12, 481.* Ann Arbor, MI.

WIGGINS, L. M. (1973) Panel Analysis: Latent Probability Models for Attitude and Behavior Processes. Amsterdam: Elsevier.

WINSHIP, C., and MARE, R. D. (1989) "Loglinear models with missing data: A latent class approach," in C. C. Clogg (ed.) Sociological Methodology 1989 (Vol. 19). Oxford: Basil Blackwell.

ZEISEL, H. (1968) Say It With Figures. New York: Harper & Row.

ABOUT THE AUTHOR

JACQUES A. HAGENAARS is Full Professor of Methodology of the Social Sciences at Tilburg University, the Netherlands. His current research interests center on categorical and longitudinal causal analysis. He is the author of LCAG, a program for loglinear modeling with latent variables (and missing data), coeditor of a Dutch textbook on causal analysis, author of *Categorical Longitudinal Data: Loglinear Panel, Trend, and Cohort Analysis* (Sage, 1990), and has written many articles that have appeared in national (Dutch) and international scholarly journals. At present he is involved in research projects concerning general loglinear modeling and causality and quasi-experimentation.

Library &
Information
Services

Aston Triangle
Birmingham
B4 7ET
England

Tel +44 (0121) 359 3611
Fax +44 (0121) 359 7358
email library@aston.ac.uk
Website http://www.lis.aston.ac.uk/